Practice

Eureka Math®

Grade 1 Fluency Modules 1–3

Published by Great Minds®.

Printed in the U.S.A.
This book may be purchased from the publisher at eureka-math.org.
CMP 10 9 8 7 6 5 4 3 2 1

ISBN 978-1-64054-566-3

G1-M1-M3-P/F-04.2018

Learn • Practice • Succeed

Eureka Math® student materials for *A Story of Units®* (K–5) are available in the *Learn, Practice, Succeed* trio. This series supports differentiation and remediation while keeping student materials organized and accessible. Educators will find that the *Learn, Practice,* and *Succeed* series also offers coherent—and therefore, more effective—resources for Response to Intervention (RTI), extra practice, and summer learning.

Learn

Eureka Math Learn serves as a student's in-class companion where they show their thinking, share what they know, and watch their knowledge build every day. *Learn* assembles the daily classwork—Application Problems, Exit Tickets, Problem Sets, templates—in an easily stored and navigated volume.

Practice

Each *Eureka Math* lesson begins with a series of energetic, joyous fluency activities, including those found in *Eureka Math Practice.* Students who are fluent in their math facts can master more material more deeply. With *Practice,* students build competence in newly acquired skills and reinforce previous learning in preparation for the next lesson.

Together, *Learn* and *Practice* provide all the print materials students will use for their core math instruction.

Succeed

Eureka Math Succeed enables students to work individually toward mastery. These additional problem sets align lesson by lesson with classroom instruction, making them ideal for use as homework or extra practice. Each problem set is accompanied by a Homework Helper, a set of worked examples that illustrate how to solve similar problems.

Teachers and tutors can use *Succeed* books from prior grade levels as curriculum-consistent tools for filling gaps in foundational knowledge. Students will thrive and progress more quickly as familiar models facilitate connections to their current grade-level content.

Students, families, and educators:

Thank you for being part of the *Eureka Math*® community, where we celebrate the joy, wonder, and thrill of mathematics. One of the most obvious ways we display our excitement is through the fluency activities provided in *Eureka Math Practice*.

What is fluency in mathematics?

You may think of *fluency* as associated with the language arts, where it refers to speaking and writing with ease. In prekindergarten through grade 5, the *Eureka Math* curriculum contains multiple daily opportunities to build fluency *in mathematics*. Each is designed with the same notion—growing every student's ability to use mathematics *with ease*. Fluency experiences are generally fast-paced and energetic, celebrating improvement and focusing on recognizing patterns and connections within the material. They are not intended to be graded.

Eureka Math fluency activities provide differentiated practice through a variety of formats—some are conducted orally, some use manipulatives, others use a personal whiteboard, and still others use a handout and paper-and-pencil format. *Eureka Math Practice* provides each student with the printed fluency exercises for his or her grade level.

What is a Sprint?

Many printed fluency activities utilize the format we call a Sprint. These exercises build speed and accuracy with already acquired skills. Used when students are nearing optimum profi ciency, Sprints leverage tempo to build a low-stakes adrenaline boost that increases memory and recall. Their intentional design makes Sprints inherently differentiated; the problems build from simple to complex, with the first quadrant of problems being the simplest and each subsequent quadrant adding complexity. Further, intentional patterns within the sequence of problems engage students' higher order thinking skills.

The suggested format for delivering a Sprint calls for students to do two consecutive Sprints (labeled A and B) on the same skill, each timed at one minute. Students pause between Sprints to articulate the patterns they noticed as they worked the fi rst Sprint. Noticing the patterns often provides a natural boost to their performance on the second Sprint.

Sprints can be conducted with an untimed protocol as well. The untimed protocol is highly recommended when students are still building confi dence with the level of complexity of the fi rst quadrant of problems. Once all students are prepared for success on the Sprint, the work of improving speed and accuracy with the energy of a timed protocol is often welcome and invigorating.

Where can I find other fluency activities?

The *Eureka Math Teacher Edition* guides educators in the delivery of all fluency activities for each lesson, including those that do not require print materials. Additionally, the *Eureka Digital Suite* provides access to the fl uency activities for all grade levels, searchable by standard or lesson.

Best wishes for a year filled with aha moments!

Jill Diniz

Jill Diniz
Director of Mathematics
Great Minds

Contents

Module 1

Module 2

Module 3

Grade 1
Module 1

A

Number Correct:

Name ______________________ Date ____________

*Write the number of dots. Find 1 or 2 groups that make finding the total number of dots easier

1.	●●		16.	●●●●● ●●●●	
2.	●●●		17.	●●●●● ●●●	
3.	●●●●		18.	●●●●● ●●●●●	
4.	●●●		19.	●●●●● ●●	
5.	●		20.	●●●●● ●	
6.	●●●●		21.	●●●●● ●●●●	
7.	●●●●●		22.	●●●●● ●●●●●	
8.	●●●●		23.	●●●● ●●●●●	
9.	●●●●● ●		24.	●●●●● ●●●	
10.	●●●●● ●●		25.	●●● ●● ●●●●●	
11.	●●●●●		26.	●●●●● ●●	
12.	●●●●		27.	●●● ●● ●● ●●●	
13.	●●●●● ●		28.	●● ●● ● ●● ●●	
14.	●●●●● ●●●		29.	●● ●●● ● ●●	
15.	●●●●● ●●		30.	●● ● ●● ●● ● ●●	

B

Number Correct:

Name ______________________ Date ____________

*Write the number of dots. Find 1 or 2 groups that make finding the total number of dots easier

1.	●		16.	●●●●● ●●●	
2.	●●		17.	●●●●● ●●●●	
3.	●		18.	●●●●● ●●	
4.	●●●●		19.	●●●●● ●●●	
5.	●●●		20.	●●●●● ●●●●●	
6.	●●●●●		21.	●●●●● ●●●●	
7.	●●●●		22.	●●●●● ●●●●●	
8.	●●●●●		23.	● ●●●● ●●●●●	
9.	●●●●● ●●		24.	●●●●● ●●●●●	
10.	●●●●● ●		25.	●● ●●●●●	
11.	●●●●● ●●●		26.	●●● ● ●● ●●	
12.	●●●●● ●		27.	●● ●●● ●●● ●●	
13.	●●●●●		28.	●● ● ● ● ●● ●●	
14.	●●●●● ●●		29.	●● ●● ● ●● ●●	
15.	●●●●● ●		30.	●● ● ●●●● ●●	

Name ______________________________ Date __________

Number Bond Dash!

Do as many as you can in 90 seconds. Write the number of bonds you finished here:

1.

2.
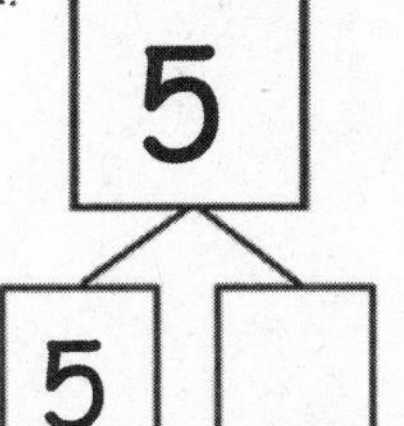

3.

4.
5
3

5.
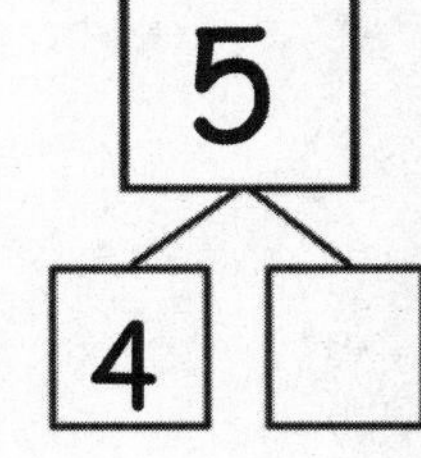

6.
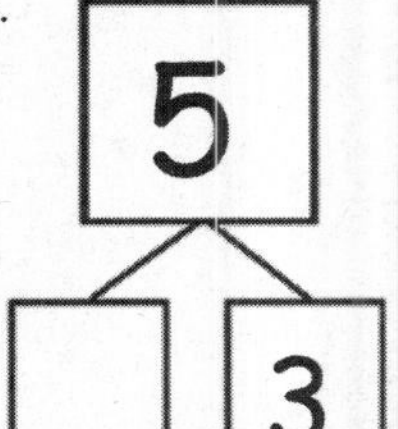

7.

8.
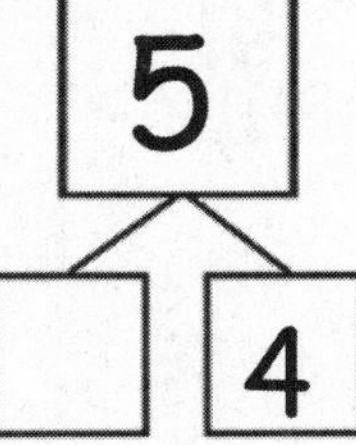

9.
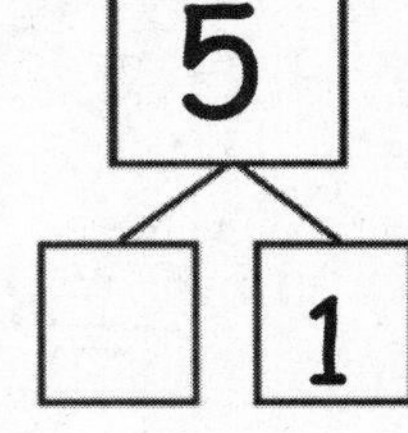

10.
5
2

11.
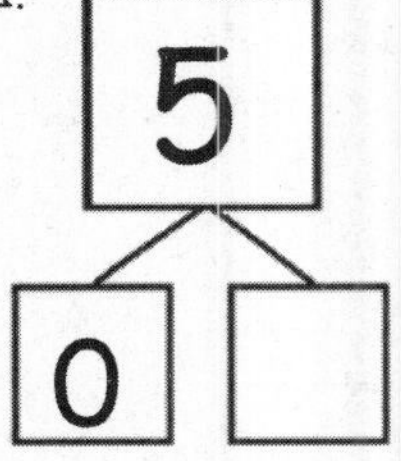

12.
5
1

13.

14.

15.
5
4

16.
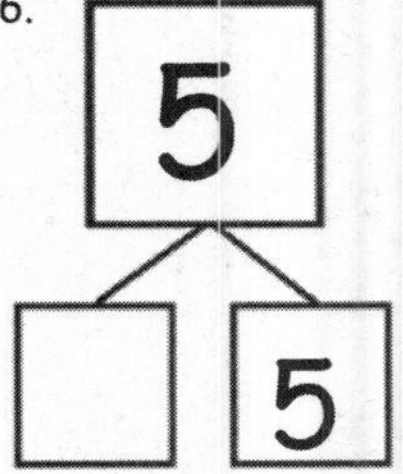

17.
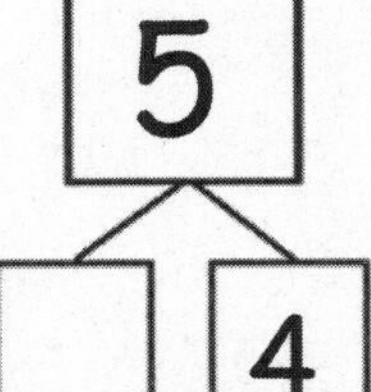

18.

19.
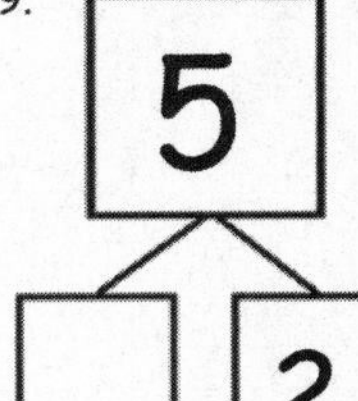

20.
5
1

21.
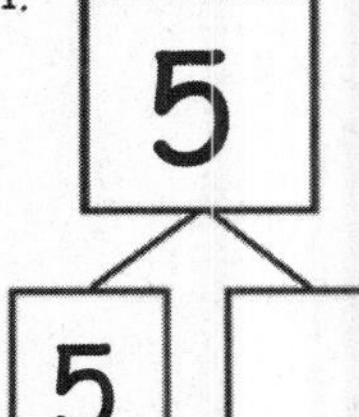

22.
5
0

23.
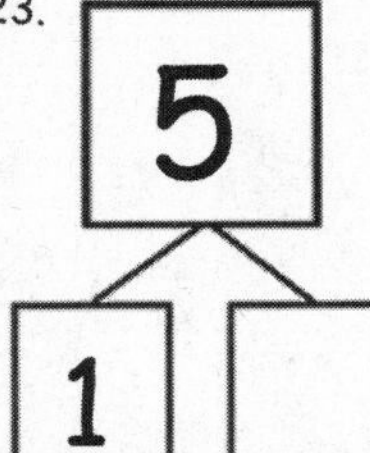

24.
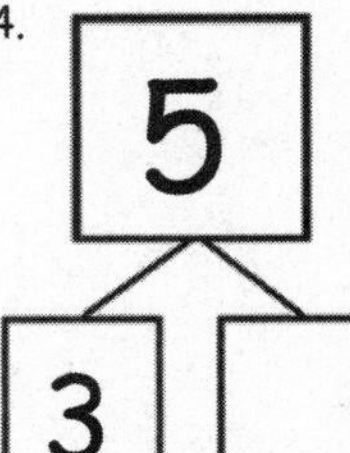

25.
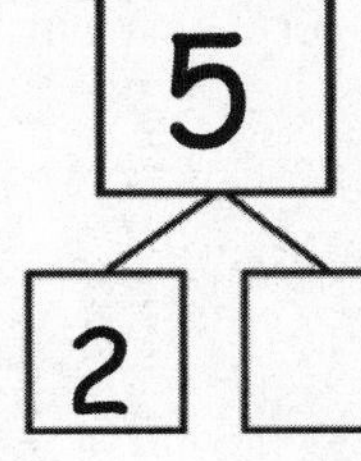

____

number bond dash 5

A

Number Correct:

Name ______________________ Date ____________

*Write the number that is 1 more.

1.	●●●		16.	●●●●● ●●●●	
2.	●●		17.	9	
3.	●●●		18.	7	
4.	●●●●		19.	●●●●● ●●	
5.	●●●●●		20.	8	
6.	●●●●● ●		21.	7	
7.	●●●●●		22.	●●●●● ●●●	
8.	5		23.	●●●●● ●●●●	
9.	●●●●● ●●		24.	10	
10.	6		25.	●●●●● ●●●●●	
11.	●●●●● ●		26.	●●●●● ●●●	
12.	7		27.	●● ●● ●● ●●	
13.	●●●●● ●●		28.	9	
14.	●●●●● ●●●		29.	●●● ●●● ●●●	
15.	8		30.	●●● ●●● ●●● ●●●	

B

Number Correct:

Name ______________________ Date ______________

*Write the number that is 1 more.

1.	●●		16.	●●●●● ●●●	
2.	●		17.	8	
3.	●●		18.	9	
4.	●●●		19.	●●●●● ●●●●	
5.	●●●●		20.	●●●●● ●●●●●	
6.	●●●●●		21.	10	
7.	●●●●		22.	●●●●● ●●●	
8.	4		23.	●●●●● ●●●●	
9.	●●●●●		24.	10	
10.	5		25.	●●●●● ●●●●	
11.	●●●●●		26.	●● ●● ● ●● ●●	
12.	7		27.	●● ●● ●● ●●	
13.	●●●●● ●●		28.	8	
14.	●●●●● ●		29.	●● ●● ●● ●●●	
15.	6		30.	●●● ●●●● ●● ●●●●	

Shake Those Disks!—6

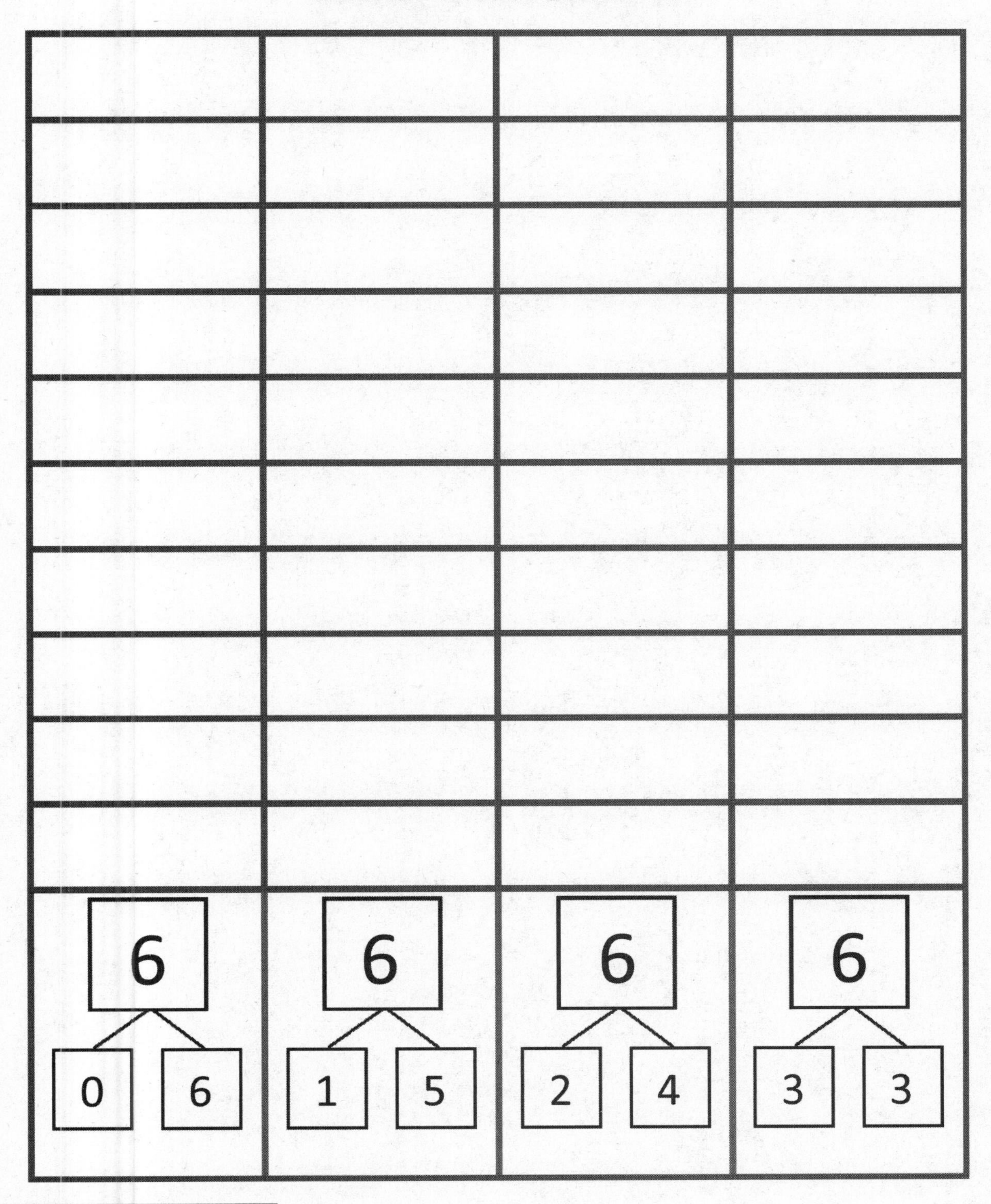

shake those disks 6 board

Name ______________________________ Date __________

Do as many as you can in 90 seconds. Write the number of bonds you finished here:

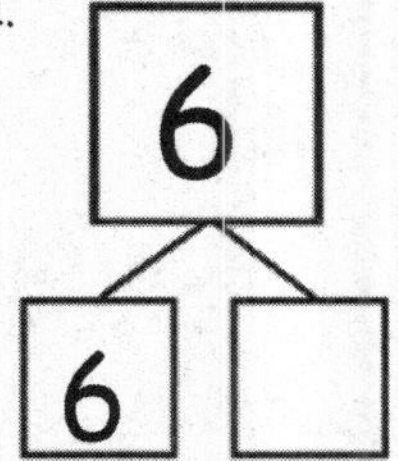
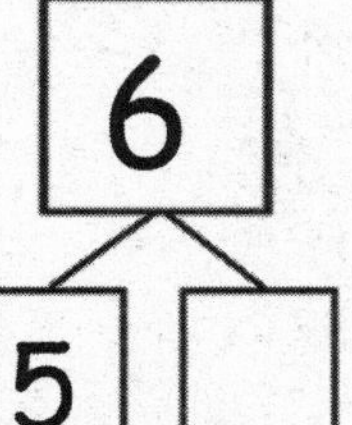

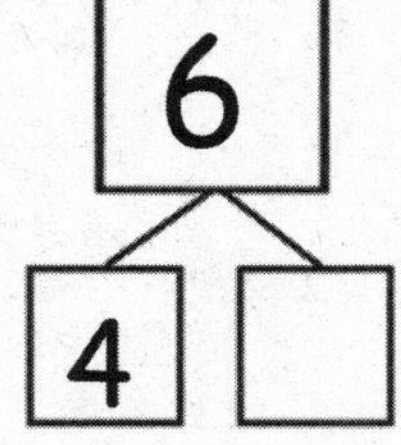
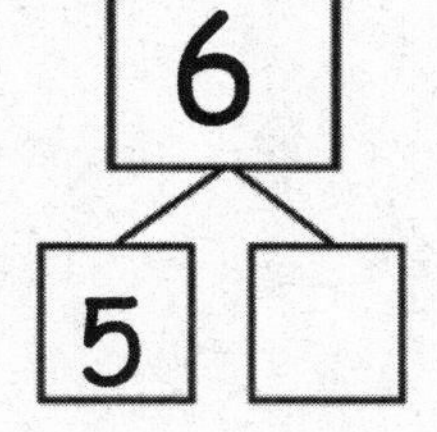
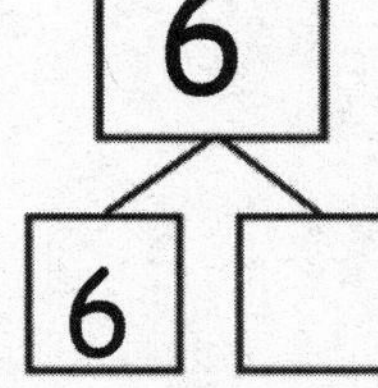

1. 6 — 6, ☐
2. 6 — 5, ☐
3. 6 — 4, ☐
4. 6 — 5, ☐
5. 6 — 6, ☐

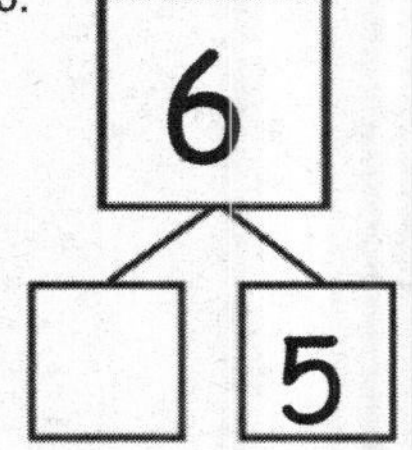
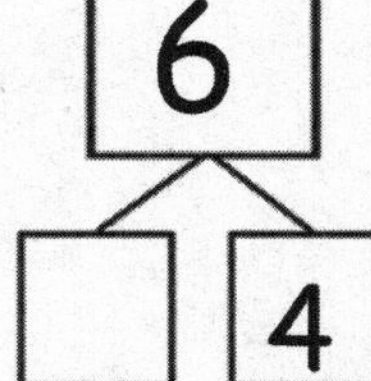

6. 6 — ☐, 5
7. 6 — ☐, 4
8. 6 — ☐, 5
9. 6 — ☐, 4
10. 6 — ☐, 3

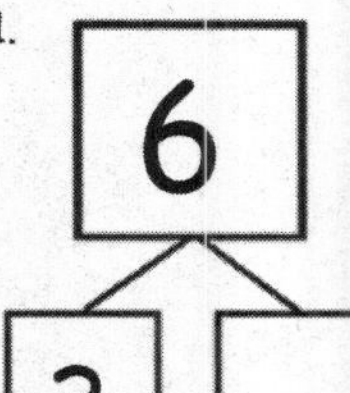

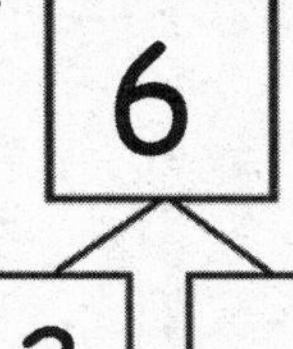
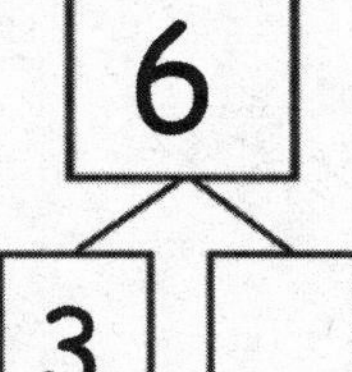
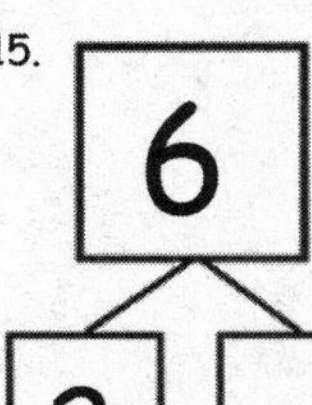

11. 6 — 3, ☐
12. 6 — 4, ☐
13. 6 — 2, ☐
14. 6 — 3, ☐
15. 6 — 2, ☐

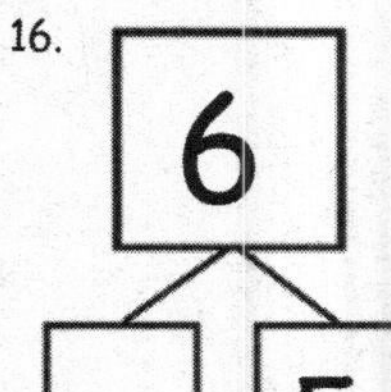

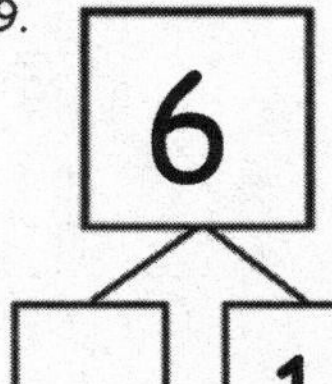
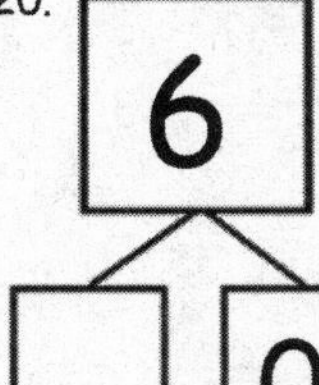

16. 6 — ☐, 5
17. 6 — ☐, 1
18. 6 — ☐, 0
19. 6 — ☐, 1
20. 6 — ☐, 0

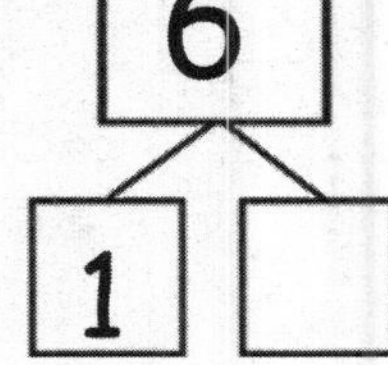

21. 6 — 1, ☐
22. 6 — 5, ☐
23. 6 — 4, ☐
24. 6 — 2, ☐
25. 6 — 3, ☐

number bond dash 6

Name ______________________________ Date __________

Do as many as you can in 90 seconds. Write the number of bonds you finished here:

1. 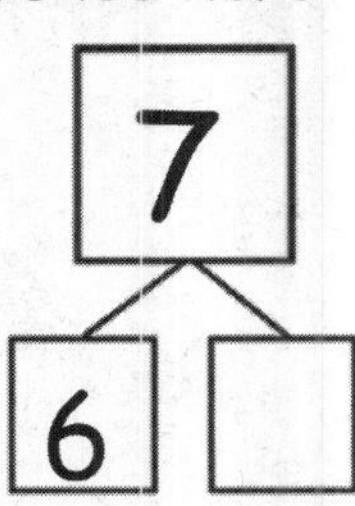
2. 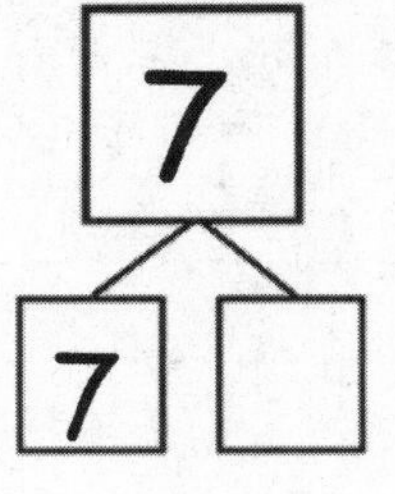
3. 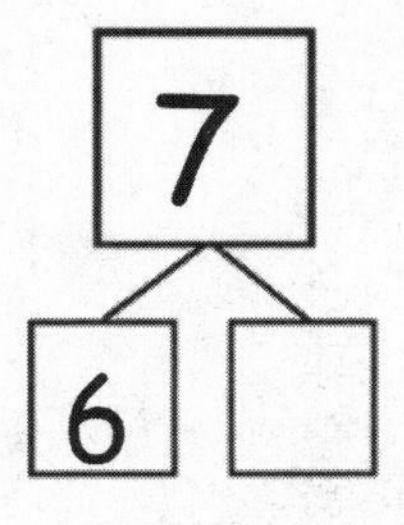
4. 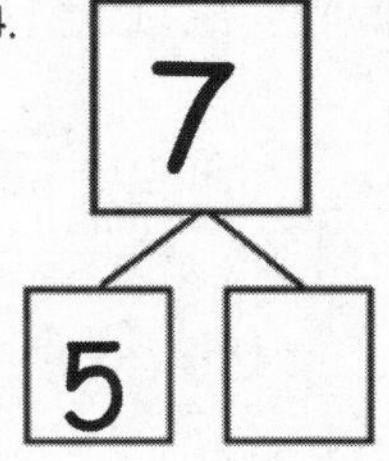
5. 
6. 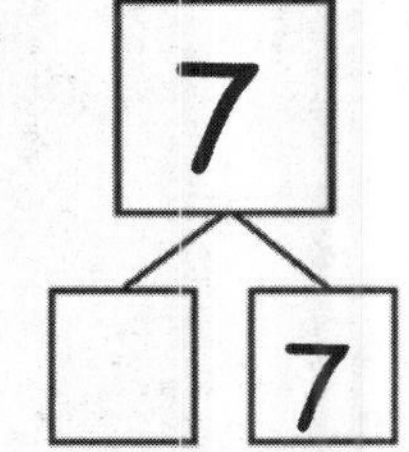
7. 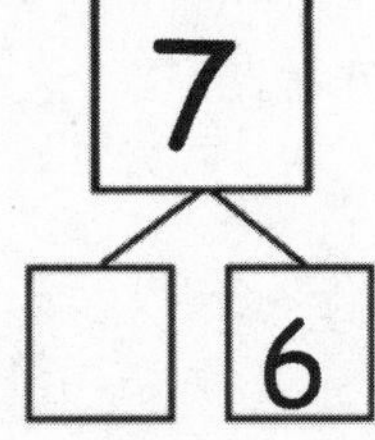
8. 
9. 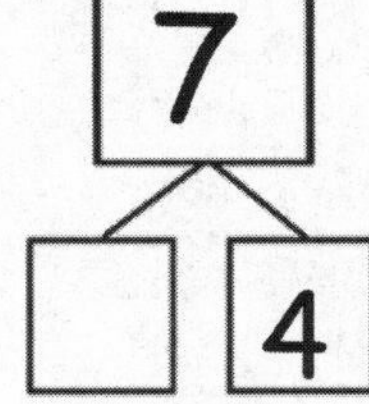
10. 
11. 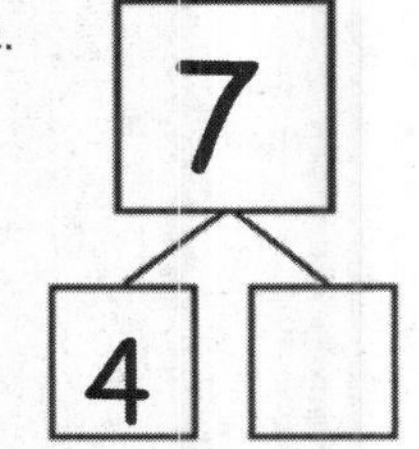
12. 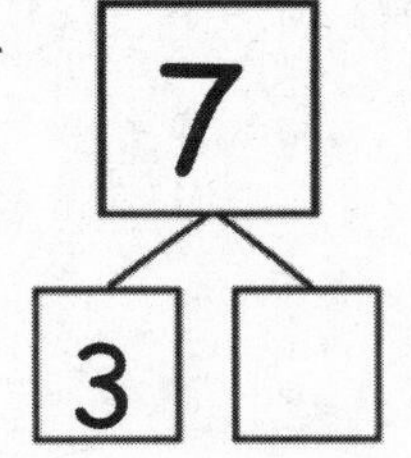
13. 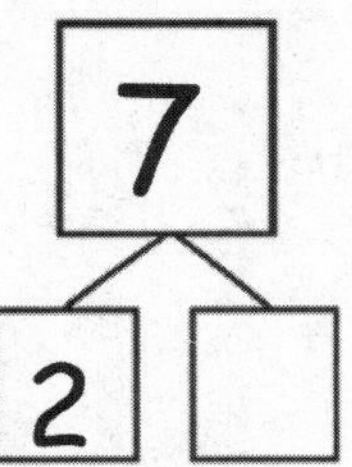
14. 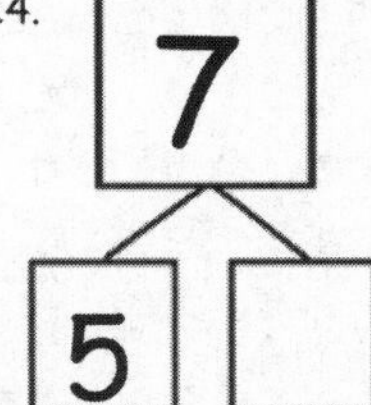
15. 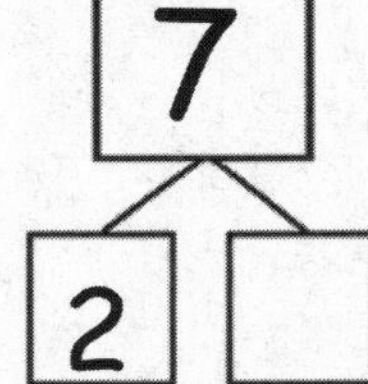
16. 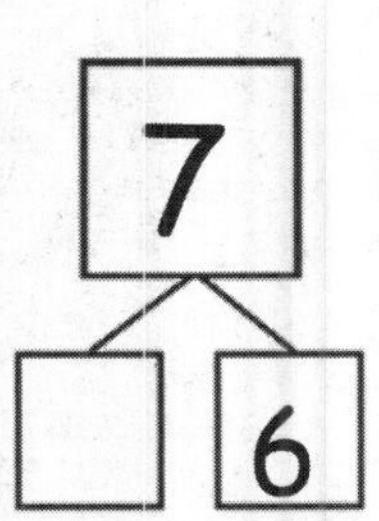
17. 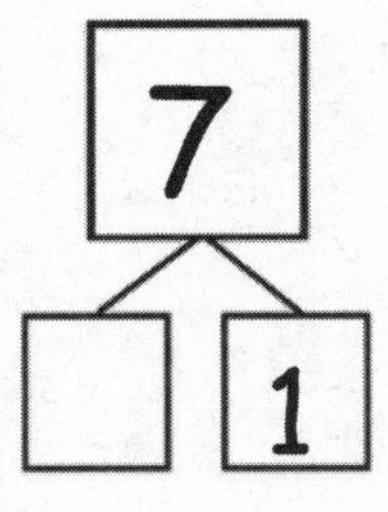
18. 
19. 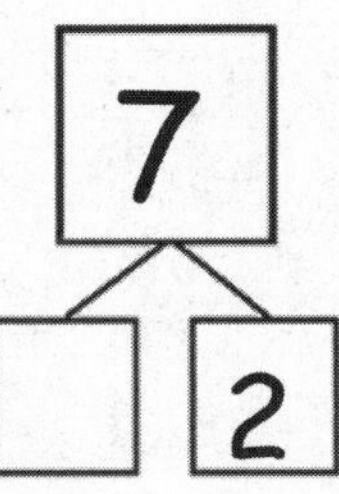
20. 7 / [ ] 5
21. 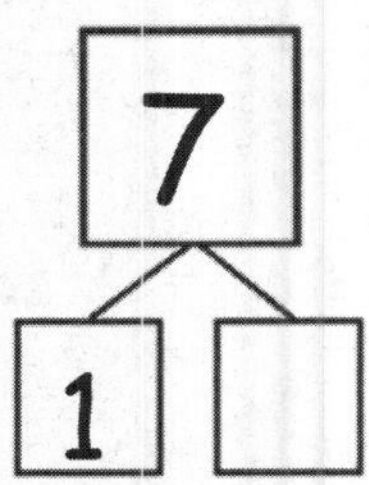
22. 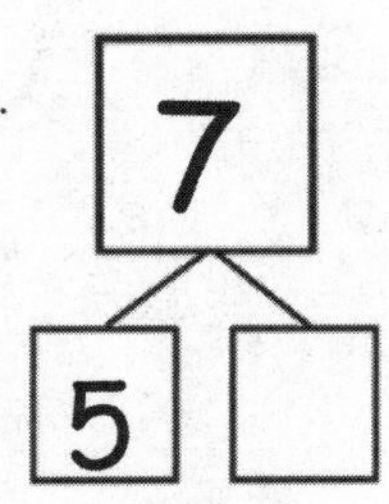
23. 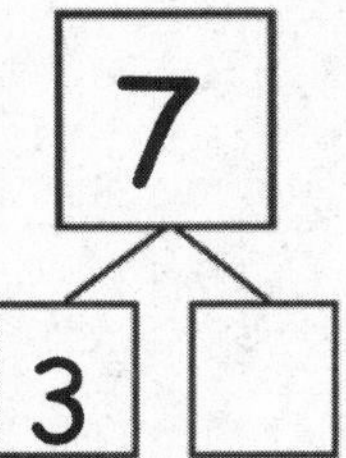
24. 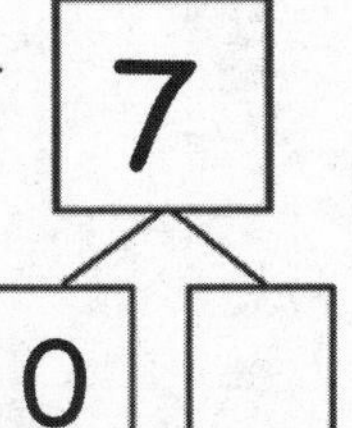
25. 

number bond dash 7

Shake Those Disks!—8

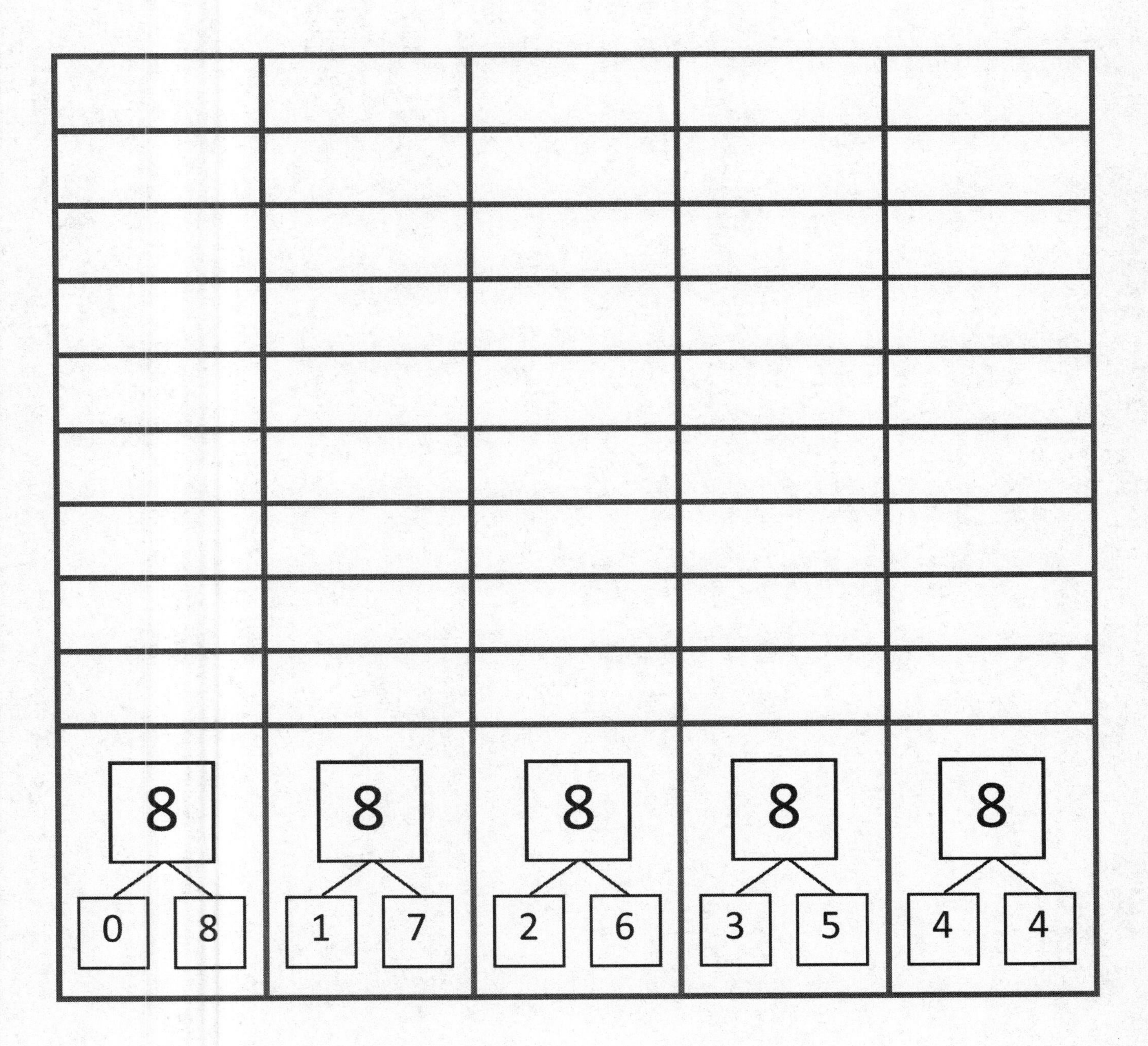

shake those disks 8

Name ______________________________ Date ____________

Do as many as you can in 90 seconds. Write the number of bonds you finished here:

1. 8: 8, ___
2. 8: 7, ___
3. 8: 6, ___
4. 8: 7, ___
5. 8: 6, ___
6. 8: ___, 5
7. 8: ___, 6
8. 8: ___, 5
9. 8: ___, 4
10. 8: ___, 3
11. 8: 4, ___
12. 8: 5, ___
13. 8: 3, ___
14. 8: 4, ___
15. 8: 3, ___
16. 8: ___, 6
17. 8: ___, 2
18. 8: ___, 6
19. 8: ___, 5
20. 8: ___, 3
21. 8: 4, ___
22. 8: 1, ___
23. 8: 2, ___
24. 8: 0, ___
25. 8: 1, ___

number bond dash 8

Name ______________________________ Date ____________

Do as many as you can in 90 seconds. Write the number of bonds you finished here:

#	Whole	Part	Part
1.	9	8	
2.	9	7	
3.	9	8	
4.	9	7	
5.	9	9	
6.	9		6
7.	9		7
8.	9		6
9.	9		5
10.	9		4
11.	9	8	
12.	9	1	
13.	9	7	
14.	9	2	
15.	9	6	
16.	9		5
17.	9		6
18.	9		7
19.	9		2
20.	9		3
21.	9	5	
22.	9	1	
23.	9	2	
24.	9	0	
25.	9	2	

number bond dash 9

Name ____________________ Date __________

Do as many as you can in 90 seconds. Write the number of bonds you finished here:

1. 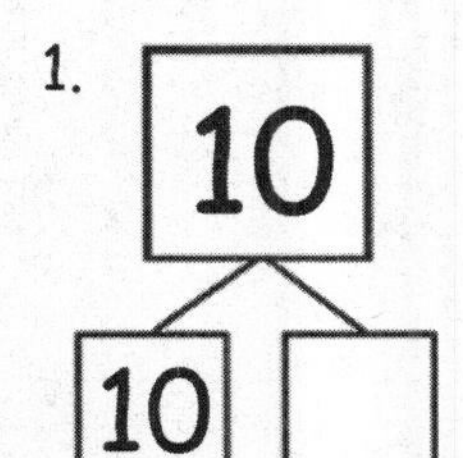

2. 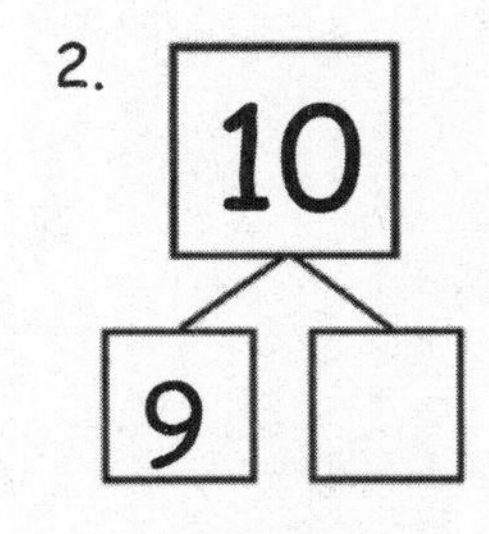

3. 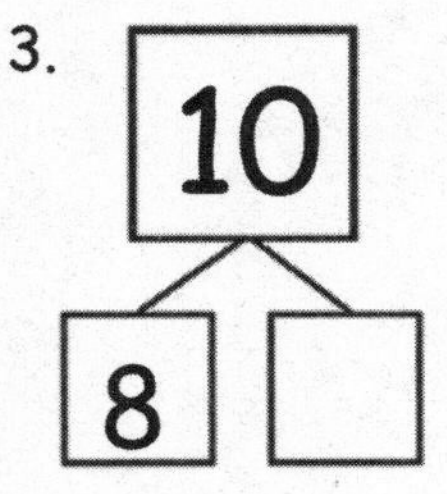

4. 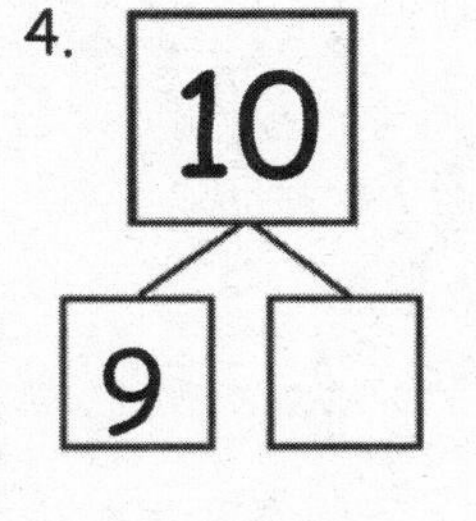

5. 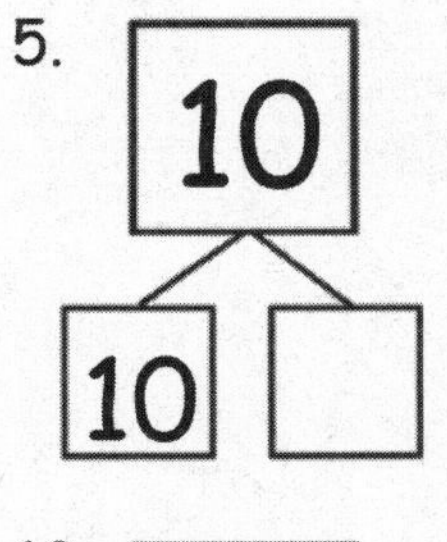

6. 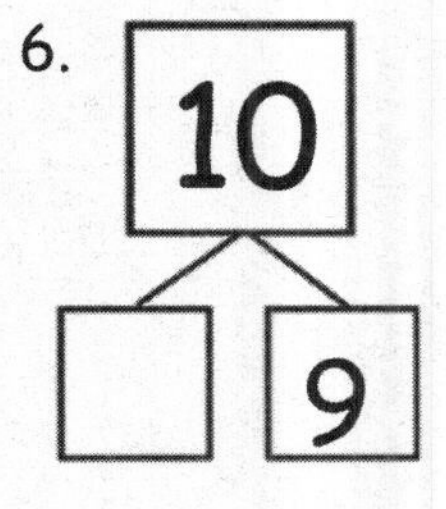

7. 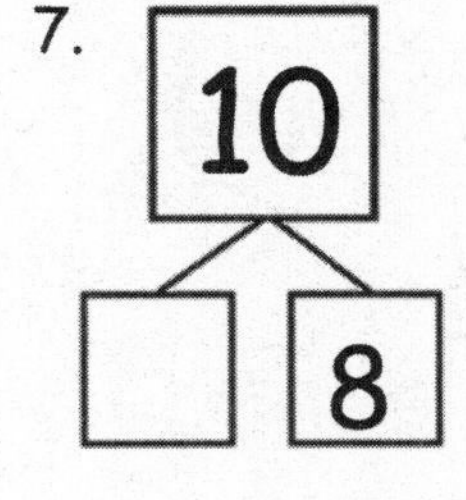

8. 

9. 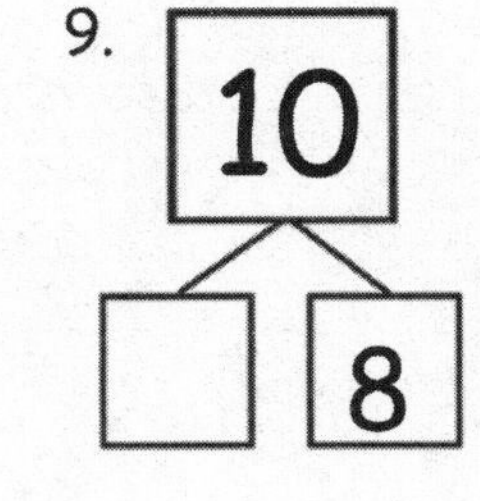

10. 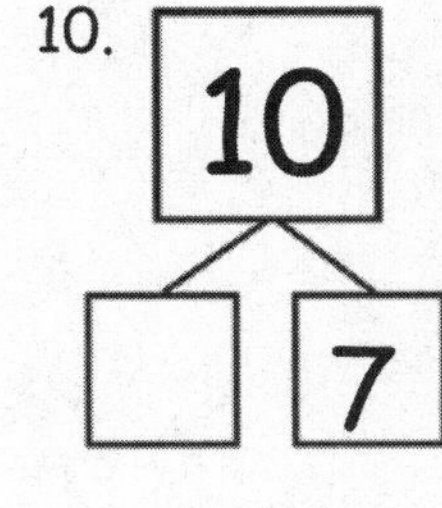

11. 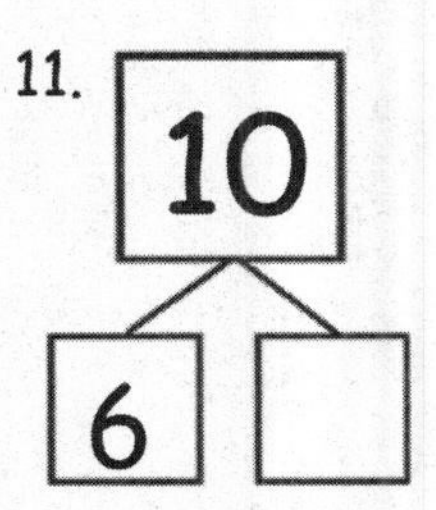

12. 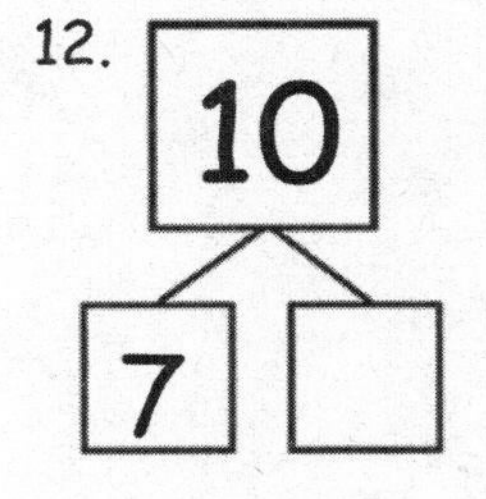

13. 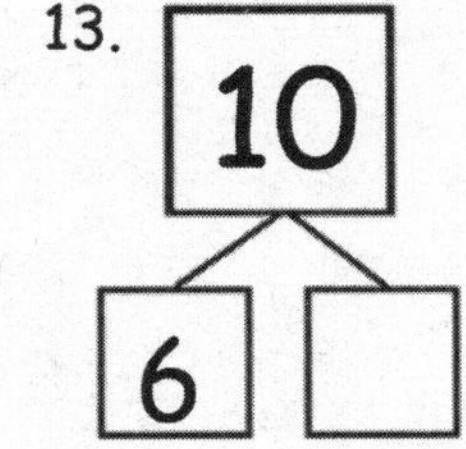

14. 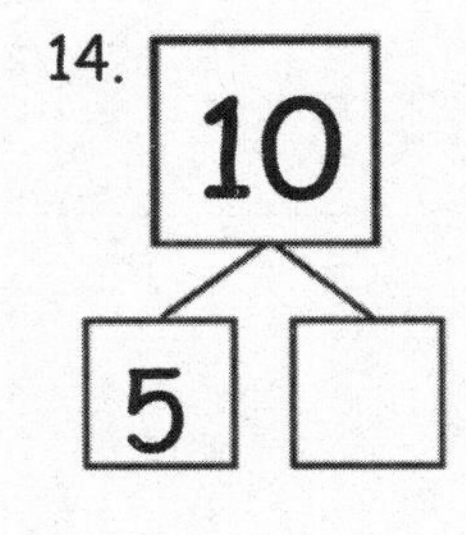

15. 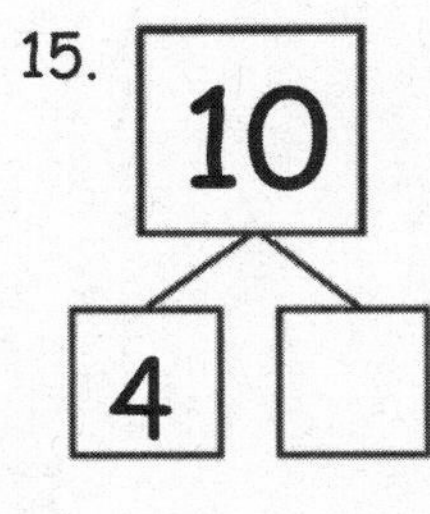

16. 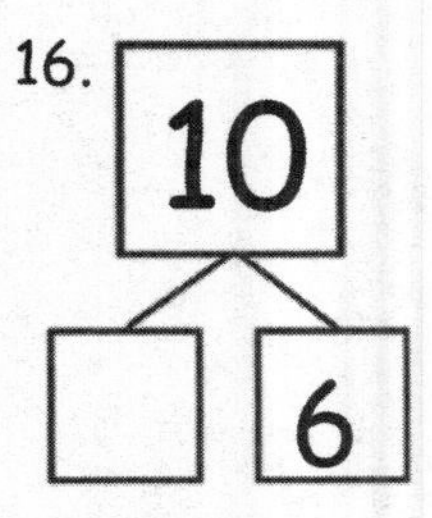

17. 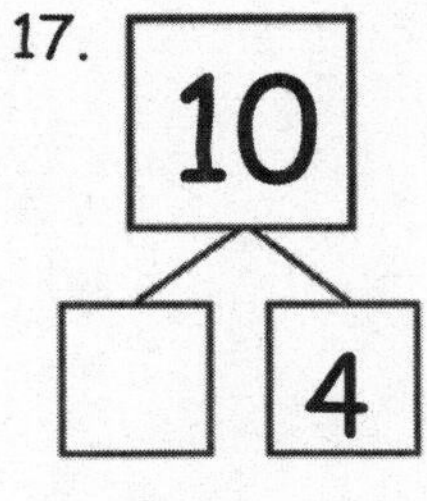

18. 

19. 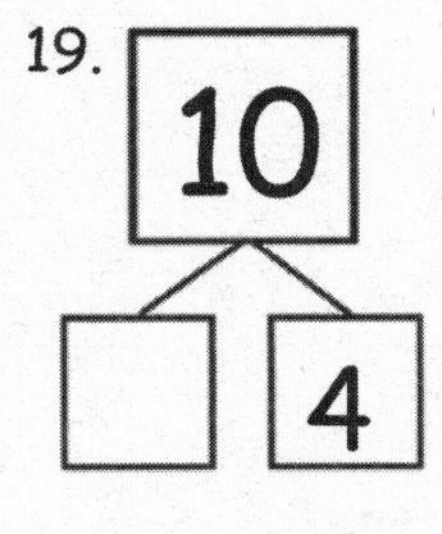

20. 

21. 10

0

22. 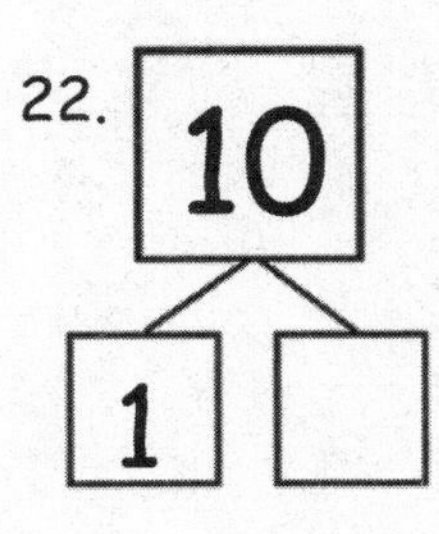

23. 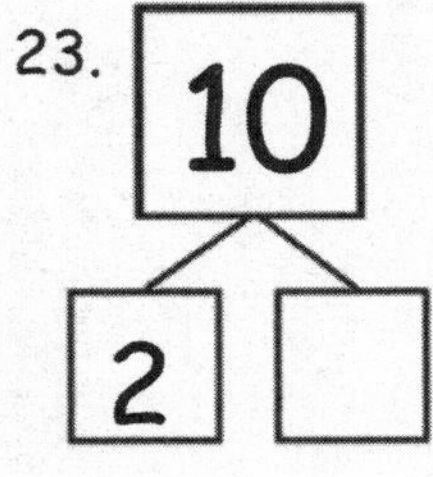

24. 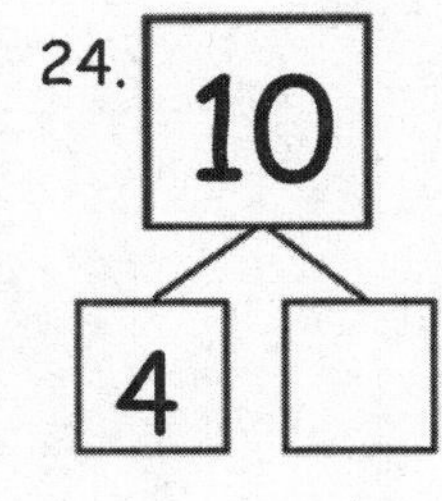

25. 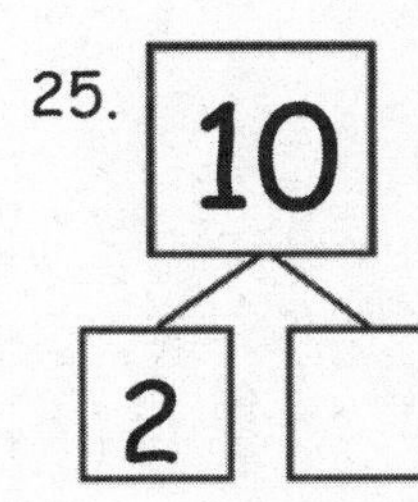

number bond dash 10

Target Number:

Target Practice

Choose a *target number* between 6 and 10 and write it in the middle of the circle on the top of the page. Roll a die. Write the number rolled in the circle at the end one of the arrows. Then, make a bull's-eye by writing the number needed to make your target in the other circle.

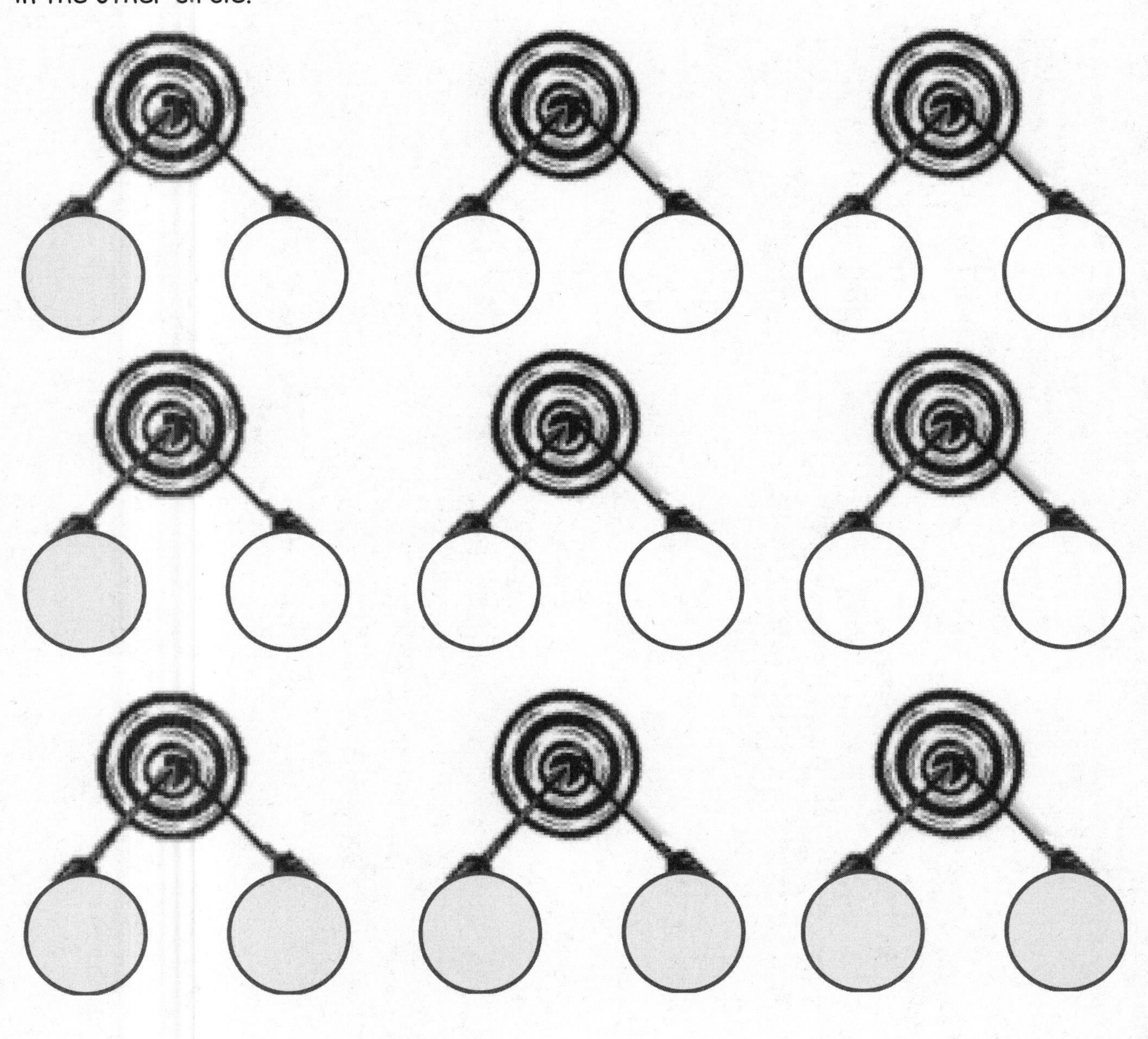

target practice

A

Number Correct:

Name ______________________ Date ______________

*Count on to add. Write the number.

1.	1 + 1 ● ●		16.	4 + 3 ●●●	
2.	2 + 1 ●● ●		17.	5 + 3 ●●●	
3.	3 + 1 ●●● ●		18.	7 + 3 ●●●	
4.	3 + 2 ●●● ●●		19.	7 + 2 ●●	
5.	1 + 2 ● ●●		20.	8 + 2 ●●	
6.	2 + 2 ●● ●●		21.	6 + 2 ●●	
7.	2 + 3 ●● ●●●		22.	6 + 1 ●	
8.	2 + 1 ●		23.	6 + 1	
9.	2 + 2 ●●		24.	6 + 2	
10.	3 + 2 ●●		25.	7 + 2	
11.	5 + 2 ●●		26.	8 + 2	
12.	8 + 2 ●●		27.	2 + 8	
13.	8 + 1 ●		28.	2 + 6	
14.	7 + 1 ●		29.	3 + 6	
15.	9 + 1 ●		30.	4 + 5	

B

Number Correct:

Name ______________________ Date ____________

*Count on to add. Write the number.

1.	1 + 1		16.	4 + 2	
2.	2 + 2		17.	3 + 2	
3.	3 + 2		18.	5 + 2	
4.	2 + 2		19.	7 + 2	
5.	2 + 1		20.	7 + 3	
6.	3 + 1		21.	6 + 3	
7.	3 + 2		22.	6 + 2	
8.	3 + 2		23.	6 + 2	
9.	2 + 2		24.	5 + 2	
10.	4 + 2		25.	7 + 2	
11.	1 + 2		26.	6 + 2	
12.	2 + 1		27.	2 + 6	
13.	3 + 1		28.	2 + 7	
14.	5 + 1		29.	3 + 7	
15.	7 + 1		30.	4 + 7	

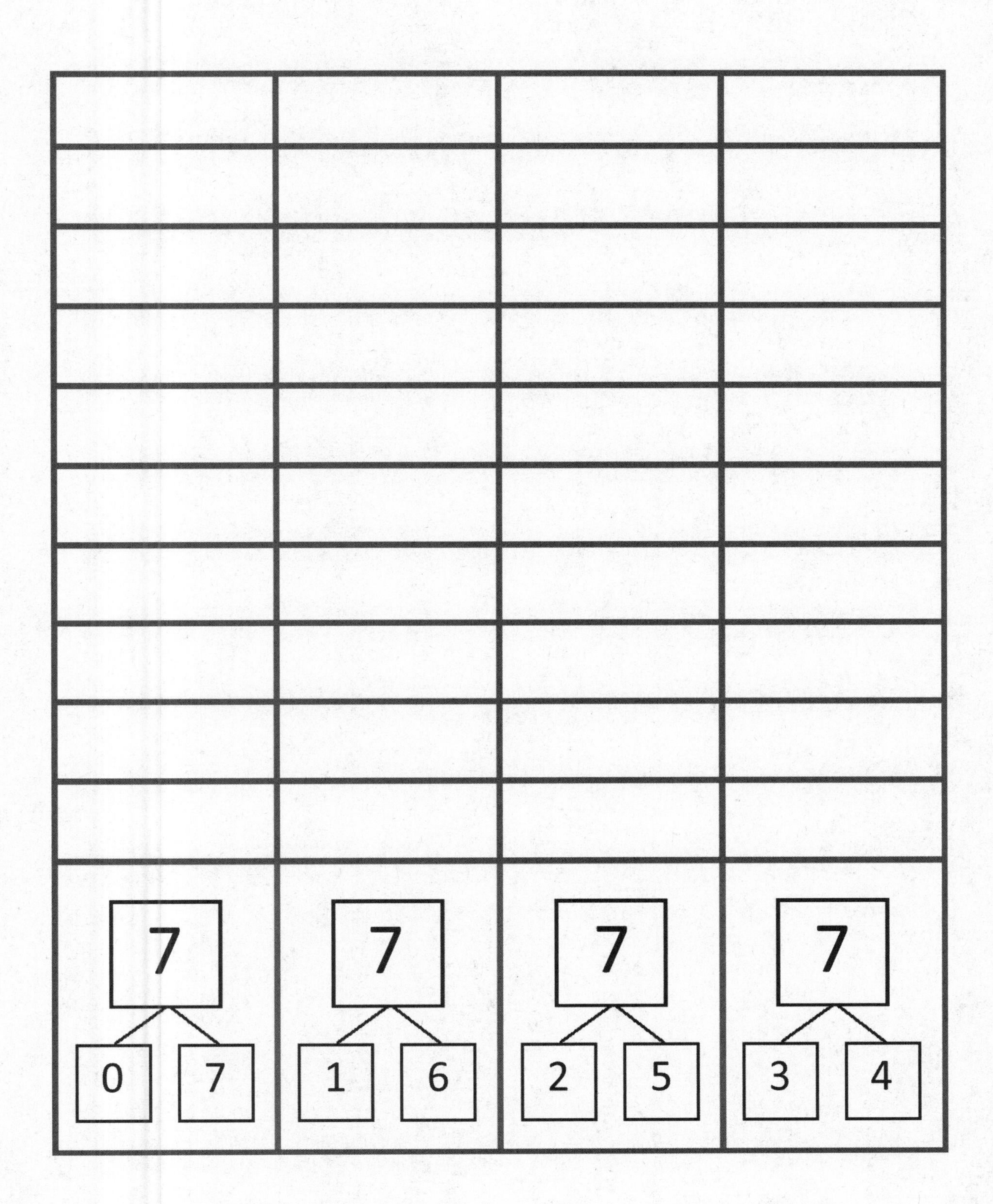

shake those disks 7 board

A

Name ______________________ Date ____________

Number Correct:

*Count on to add.

1.	1 + 1		16.	4 + 3	
2.	2 + 1		17.	3 + 3	
3.	3 + 1		18.	4 + 3	
4.	3 + 2		19.	3 + 4	
5.	2 + 2		20.	2 + 4	
6.	3 + 2		21.	4 + 2	
7.	2 + 2		22.	5 + 2	
8.	3 + 0		23.	2 + 5	
9.	3 + 1		24.	2 + 6	
10.	3 + 2		25.	6 + 3	
11.	5 + 2		26.	3 + 6	
12.	5 + 3		27.	2 + 7	
13.	5 + 2		28.	3 + 7	
14.	5 + 3		29.	2 + 8	
15.	6 + 3		30.	3 + 6	

B

Number Correct:

Name ______________________ Date ____________

*Count on to add.

1.	2 + 1		16.	4 + 3	
2.	1 + 1		17.	3 + 3	
3.	2 + 1		18.	2 + 3	
4.	2 + 2		19.	1 + 3	
5.	3 + 2		20.	0 + 3	
6.	2 + 2		21.	1 + 3	
7.	3 + 2		22.	2 + 5	
8.	3 + 1		23.	5 + 2	
9.	5 + 1		24.	2 + 6	
10.	6 + 1		25.	6 + 2	
11.	6 + 2		26.	3 + 6	
12.	5 + 2		27.	3 + 7	
13.	6 + 2		28.	2 + 7	
14.	6 + 3		29.	2 + 6	
15.	5 + 3		30.	3 + 6	

Name ______________________________ Date ________________

Race to the Top!

0	**2**	**4**	**6**	**8**	**10**

A

Number Correct:

Name ______________________ Date ____________

*Write the number that is 1 less.

1.	5		16.	10	
2.	4		17.	8	
3.	3		18.	11	
4.	5		19.	10	
5.	3		20.	9	
6.	1		21.	1	
7.	4		22.	11	
8.	5		23.	21	
9.	7		24.	4	
10.	6		25.	14	
11.	7		26.	24	
12.	9		27.	10	
13.	8		28.	20	
14.	9		29.	21	
15.	10		30.	31	

B

Number Correct:

Name ____________________ Date __________

*Write the number that is 1 less.

1.	3		16.	10	
2.	2		17.	9	
3.	1		18.	11	
4.	6		19.	9	
5.	4		20.	13	
6.	2		21.	11	
7.	1		22.	1	
8.	3		23.	11	
9.	5		24.	21	
10.	7		25.	5	
11.	10		26.	15	
12.	9		27.	25	
13.	8		28.	20	
14.	6		29.	10	
15.	17		30.	21	

A

Number Correct: ______

Addition

1.	3 + 1 =	
2.	4 + 1 =	
3.	5 + 1 =	
4.	9 + 1 =	
5.	6 + 1 =	
6.	8 + 1 =	
7.	2 + 1 =	
8.	7 + 1 =	
9.	1 + 7 =	
10.	1 + 9 =	
11.	1 + 6 =	
12.	2 + 2 =	
13.	3 + 2 =	
14.	4 + 2 =	
15.	8 + 2 =	
16.	5 + 2 =	
17.	6 + 2 =	
18.	7 + 2 =	
19.	2 + 7 =	
20.	2 + 8 =	
21.	2 + 5 =	
22.	2 + 6 =	

23.	1 + 2 =	
24.	3 + 6 =	
25.	1 + 8 =	
26.	2 + 3 =	
27.	1 + 4 =	
28.	2 + 4 =	
29.	1 + 3 =	
30.	1 + 5 =	
31.	3 + 3 =	
32.	4 + 3 =	
33.	5 + 3 =	
34.	6 + 3 =	
35.	7 + 3 =	
36.	3 + 7 =	
37.	3 + 4 =	
38.	3 + 5 =	
39.	4 + 4 =	
40.	5 + 4 =	
41.	6 + 4 =	
42.	4 + 6 =	
43.	4 + 5 =	
44.	5 + 5 =	

B

Number Correct: ______

Improvement: ______

Addition

1.	2 + 1 =	
2.	3 + 1 =	
3.	4 + 1 =	
4.	8 + 1 =	
5.	5 + 1 =	
6.	7 + 1 =	
7.	9 + 1 =	
8.	6 + 1 =	
9.	1 + 6 =	
10.	1 + 9 =	
11.	1 + 7 =	
12.	2 + 2 =	
13.	3 + 2 =	
14.	4 + 2 =	
15.	7 + 2 =	
16.	5 + 2 =	
17.	8 + 2 =	
18.	6 + 2 =	
19.	2 + 6 =	
20.	2 + 8 =	
21.	2 + 5 =	
22.	2 + 7 =	

23.	1 + 8 =	
24.	3 + 7 =	
25.	1 + 5 =	
26.	2 + 4 =	
27.	1 + 4 =	
28.	2 + 3 =	
29.	1 + 3 =	
30.	1 + 2 =	
31.	3 + 3 =	
32.	4 + 3 =	
33.	5 + 3 =	
34.	7 + 3 =	
35.	6 + 3 =	
36.	3 + 6 =	
37.	3 + 5 =	
38.	3 + 4 =	
39.	4 + 4 =	
40.	5 + 4 =	
41.	6 + 4 =	
42.	4 + 6 =	
43.	4 + 5 =	
44.	5 + 5 =	

A

Number Correct:

Name ______________________ Date ______________

*Write the missing number from each subtraction sentence. Pay attention to the = sign.

1.	2 - 1 = □		16.	□ = 10 - 0	
2.	1 - 1 = □		17.	□ = 10 - 1	
3.	1 - 0 = □		18.	□ = 9 - 1	
4.	3 - 1 = □		19.	□ = 7 - 1	
5.	3 - 0 = □		20.	□ = 6 - 1	
6.	4 - 0 = □		21.	□ = 6 - 0	
7.	4 - 1 = □		22.	□ = 8 - 0	
8.	5 - 1 = □		23.	8 - □ = 8	
9.	6 - 1 = □		24.	□ - 0 = 8	
10.	6 - 0 = □		25.	7 - □ = 6	
11.	8 - 0 = □		26.	7 = 7 - □	
12.	10 - 0 = □		27.	9 = 9 - □	
13.	9 - 0 = □		28.	□ - 1 = 7	
14.	9 - 1 = □		29.	□ - 0 = 8	
15.	10 - 1 = □		30.	9 = □ - 1	

B

Number Correct:

Name ____________________ Date ____________

*Write the missing number from each subtraction sentence. Pay attention to the = sign.

1.	3 - 1 = □		16.	□ = 10 - 1	
2.	2 - 1 = □		17.	□ = 9 - 1	
3.	1 - 1 = □		18.	□ = 7 - 1	
4.	1 - 0 = □		19.	□ = 7 - 0	
5.	2 - 0 = □		20.	□ = 8 - 0	
6.	4 - 0 = □		21.	□ = 10 - 0	
7.	5 - 1 = □		22.	□ = 9 - 1	
8.	7 - 1 = □		23.	9 - □ = 8	
9.	8 - 1 = □		24.	□ - 1 = 8	
10.	9 - 0 = □		25.	7 - □ = 6	
11.	10 - 0 = □		26.	6 = 7 - □	
12.	7 - 0 = □		27.	9 = 9 - □	
13.	8 - 0 = □		28.	□ - 0 = 9	
14.	10 - 1 = □		29.	□ - 0 = 10	
15.	9 - 1 = □		30.	8 = □ - 1	

A

Number Correct:

Name ______________________ Date ____________

Write the missing number for each subtraction sentence. Pay attention to the = sign.

1.	$2 - 2 = \square$		16.	$0 = 10 - \square$	
2.	$1 - 1 = \square$		17.	$0 = 9 - \square$	
3.	$1 - 0 = \square$		18.	$0 = 8 - \square$	
4.	$3 - 3 = \square$		19.	$0 = 6 - \square$	
5.	$3 - 2 = \square$		20.	$1 = 6 - \square$	
6.	$4 - 4 = \square$		21.	$1 = 7 - \square$	
7.	$4 - 3 = \square$		22.	$1 = 10 - \square$	
8.	$6 - 6 = \square$		23.	$10 - \square = 1$	
9.	$7 - 7 = \square$		24.	$\square - 9 = 1$	
10.	$8 - 8 = \square$		25.	$7 - \square = 0$	
11.	$8 - 7 = \square$		26.	$0 = 7 - \square$	
12.	$9 - 9 = \square$		27.	$0 = 9 - \square$	
13.	$9 - 8 = \square$		28.	$\square - 8 = 0$	
14.	$10 - 10 = \square$		29.	$\square - 7 = 1$	
15.	$10 - 9 = \square$		30.	$1 = \square - 5$	

B

Number Correct:

Name ______________________ Date ____________

Write the missing number for each subtraction sentence. Pay attention to the = sign.

1.	3 - 3 = □		16.	0 = 6 - □	
2.	2 - 2 = □		17.	0 = 7 - □	
3.	1 - 1 = □		18.	0 = 8 - □	
4.	1 - 0 = □		19.	0 = 10 - □	
5.	2 - 1 = □		20.	1 = 10 - □	
6.	4 - 3 = □		21.	1 = 9 - □	
7.	5 - 4 = □		22.	1 = 7 - □	
8.	7 - 7 = □		23.	7 - □ = 1	
9.	8 - 8 = □		24.	□ - 6 = 1	
10.	9 - 9 = □		25.	6 - □ = 0	
11.	10 - 10 = □		26.	0 = 6 - □	
12.	10 - 9 = □		27.	0 = 8 - □	
13.	8 - 7 = □		28.	□ - 8 = 0	
14.	6 - 5 = □		29.	□ - 6 = 1	
15.	6 - 6 = □		30.	1 = □ - 6	

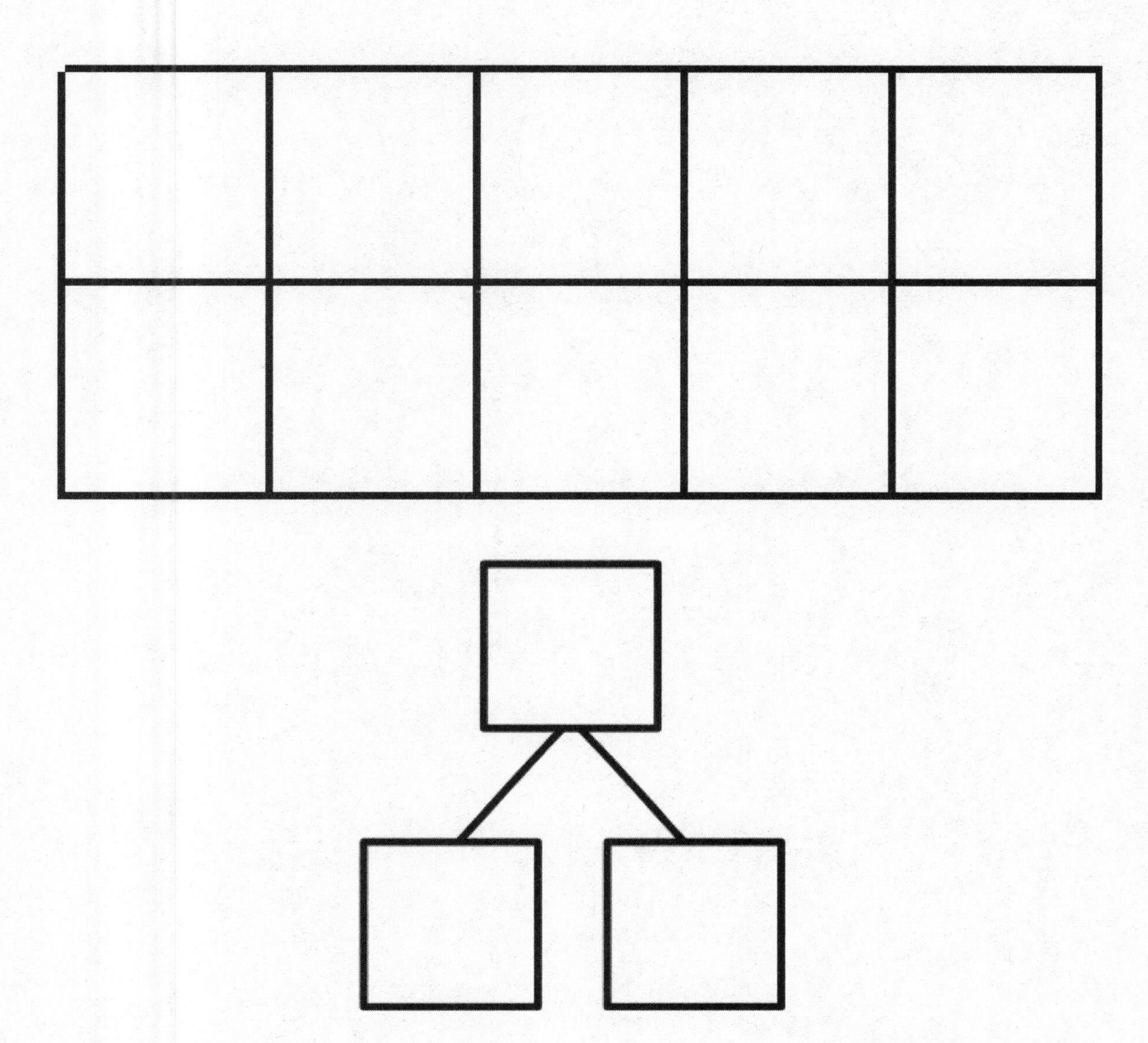

ten-frame

A

Number Correct:

Name ____________________ Date ____________

*Write the missing number for each number sentence. Pay attention to the + and - signs.

1.	9 + 1 = □		16.	10 - 7 = □	
2.	1 + 9 = □		17.	10 = 7 + □	
3.	10 - 1 = □		18.	10 = 3 + □	
4.	10 - 9 = □		19.	10 = 6 + □	
5.	10 + 0 = □		20.	10 = 4 + □	
6.	0 + 10 = □		21.	10 = 5 + □	
7.	10 - 0 = □		22.	10 - □ = 5	
8.	10 - 10 = □		23.	5 = 10 - □	
9.	8 + 2 = □		24.	6 = 10 - □	
10.	2 + 8 = □		25.	7 = 10 - □	
11.	10 - 2 = □		26.	7 = □ - 3	
12.	10 - 8 = □		27.	4 = 10 - □	
13.	7 + 3 = □		28.	5 = □ - 5	
14.	3 + 7 = □		29.	6 = 10 - □	
15.	10 - 3 = □		30.	7 = □ - 3	

B

Number Correct:

Name ______________________ Date ______________

*Write the missing number for each number sentence. Pay attention to the + and – signs.

1.	8 + 2 = □		16.	10 - 6 = □	
2.	2 + 8 = □		17.	10 = 8 + □	
3.	10 - 2 = □		18.	10 = 7 + □	
4.	10 - 8 = □		19.	10 = 3 + □	
5.	9 + 1 = □		20.	10 = 4 + □	
6.	1 + 9 = □		21.	10 = 5 + □	
7.	10 - 1 = □		22.	10 - □ = 5	
8.	10 - 9 = □		23.	6 = 10 - □	
9.	10 + 0 = □		24.	7 = 10 - □	
10.	0 + 10 = □		25.	8 = 10 - □	
11.	10 - 0 = □		26.	7 = □ - 3	
12.	10 - 10 = □		27.	2 = 10 - □	
13.	6 + 4 = □		28.	4 = □ - 6	
14.	4 + 6 = □		29.	3 = 10 - □	
15.	10 - 4 = □		30.	7 = □ - 3	

A

Number Correct:

Name ______________________ Date ____________

*Write the missing number for each sentence.

1.	8 and 2 make ☐		16.	11 is 10 and ☐	
2.	9 and 1 make ☐		17.	11 is 1 and ☐	
3.	7 and 3 make ☐		18.	12 is 2 and ☐	
4.	6 and ☐ make 10		19.	11 is ☐ and 1	
5.	4 and ☐ make 10		20.	14 is 10 and ☐	
6.	5 and ☐ make 10		21.	15 is 5 and ☐	
7.	☐ and 5 make 10		22.	18 is 8 and ☐	
8.	13 is 10 and ☐		23.	20 is 10 and ☐	
9.	14 is 10 and ☐		24.	2 more than 10 is ☐	
10.	16 is 10 and ☐		25.	10 more than 2 is ☐	
11.	17 is 10 and ☐		26.	10 is ☐ less than 12	
12.	19 is 10 and ☐		27.	10 is ☐ less than 12	
13.	18 is 10 and ☐		28.	8 less than 18 is ☐	
14.	12 is 10 and ☐		29.	6 less than 16 is ☐	
15.	13 is 10 and ☐		30.	10 less than 20 is ☐	

B

Number Correct:

Name ______________________ Date ____________

*Write the missing number for each sentence.

1.	9 and 1 make ☐		16.	13 is 10 and ☐	
2.	8 and 2 make ☐		17.	13 is 3 and ☐	
3.	6 and 4 make ☐		18.	11 is 1 and ☐	
4.	7 and ☐ make 10		19.	11 is ☐ and 1	
5.	3 and ☐ make 10		20.	15 is ☐ and 10	
6.	4 and ☐ make 10		21.	14 is 4 and ☐	
7.	☐ and 5 make 10		22.	19 is 9 and ☐	
8.	14 is 10 and ☐		23.	20 is 10 and ☐	
9.	13 is 10 and ☐		24.	1 more than 10 is ☐	
10.	17 is 10 and ☐		25.	10 more than 1 is ☐	
11.	16 is 10 and ☐		26.	10 is ☐ less than 11	
12.	15 is 10 and ☐		27.	10 is ☐ less than 14	
13.	19 is 10 and ☐		28.	7 less than 18 is ☐	
14.	11 is 10 and ☐		29.	7 less than 16 is ☐	
15.	12 is 10 and ☐		30.	10 less than 20 is ☐	

Grade 1
Module 2

A

Number Correct:

Name ______________________ Date ____________

*Make a ten to add.

1.	9 + 1 + 3 = □		16.	6 + 4 + 5 = □	
2.	9 + 1 + 5 = □		17.	6 + 4 + 6 = □	
3.	1 + 9 + 5 = □		18.	4 + 6 + 6 = □	
4.	1 + 9 + 1 = □		19.	4 + 6 + 5 = □	
5.	5 + 5 + 4 = □		20.	4 + 5 + 6 = □	
6.	5 + 5 + 6 = □		21.	5 + 3 + 5 = □	
7.	5 + 5 + 5 = □		22.	6 + 5 + 5 = □	
8.	8 + 2 + 1 = □		23.	1 + 4 + 9 = □	
9.	8 + 2 + 3 = □		24.	9 + 1 + □ = 14	
10.	8 + 2 + 7 = □		25.	8 + 2 + □ = 11	
11.	2 + 8 + 7 = □		26.	□ + 3 + 4 = 13	
12.	7 + 3 + 3 = □		27.	2 + □ + 6 = 16	
13.	7 + 3 + 6 = □		28.	1 + 1 + □ = 11	
14.	7 + 3 + 7 = □		29.	19 = 5 + □ + 9	
15.	3 + 7 + 7 = □		30.	18 = 2 + □ + 6	

B

Number Correct:

Name ______________________ Date ____________

*Make a ten to add.

1.	5 + 5 + 4 = □		16.	6 + 4 + 2 = □	
2.	5 + 5 + 6 = □		17.	6 + 4 + 3 = □	
3.	5 + 5 + 5 = □		18.	4 + 6 + 3 = □	
4.	9 + 1 + 1 = □		19.	4 + 6 + 6 = □	
5.	9 + 1 + 2 = □		20.	4 + 7 + 6 = □	
6.	9 + 1 + 5 = □		21.	5 + 4 + 5 = □	
7.	1 + 9 + 5 = □		22.	8 + 5 + 5 = □	
8.	1 + 9 + 6 = □		23.	1 + 7 + 9 = □	
9.	8 + 2 + 4 = □		24.	9 + 1 + □ = 11	
10.	8 + 2 + 7 = □		25.	8 + 2 + □ = 12	
11.	2 + 8 + 7 = □		26.	□ + 3 + 4 = 14	
12.	7 + 3 + 7 = □		27.	3 + □ + 7 = 20	
13.	7 + 3 + 8 = □		28.	7 + 8 + □ = 17	
14.	7 + 3 + 9 = □		29.	16 = 3 + □ + 6	
15.	3 + 7 + 9 = □		30.	19 = 2 + □ + 7	

A

Number Correct:

Name ______________________ Date ____________

*Write the missing number.

1.	9 + 1 = □		16.	9 + 5 = □	
2.	10 + 1 = □		17.	9 + 6 = □	
3.	9 + 2 = □		18.	6 + 9 = □	
4.	9 + 1 = □		19.	9 + 4 = □	
5.	10 + 2 = □		20.	4 + 9 = □	
6.	9 + 3 = □		21.	9 + 8 = □	
7.	9 + 1 = □		22.	9 + 9 = □	
8.	10 + 4 = □		23.	9 + □ = 18	
9.	9 + 5 = □		24.	□ + 6 = 15	
10.	9 + 1 = □		25.	□ + 6 = 16	
11.	10 + 6 = □		26.	13 = 9 + □	
12.	9 + 7 = □		27.	17 = 8 + □	
13.	9 + 1 = □		28.	10 + 2 = 9 + □	
14.	10 + 8 = □		29.	9 + 5 = 10 + □	
15.	9 + 9 = □		30.	□ + 7 = 8 + 9	

B

Number Correct:

Name ________________________ Date ____________

*Write the missing number.

1.	9 + 1 = □		16.	5 + 9 = □	
2.	10 + 2 = □		17.	6 + 9 = □	
3.	9 + 3 = □		18.	9 + 6 = □	
4.	9 + 1 = □		19.	9 + 7 = □	
5.	10 + 1 = □		20.	7 + 9 = □	
6.	9 + 2 = □		21.	9 + 8 = □	
7.	9 + 1 = □		22.	9 + 9 = □	
8.	10 + 3 = □		23.	9 + □ = 17	
9.	9 + 4 = □		24.	□ + 5 = 14	
10.	9 + 1 = □		25.	□ + 4 = 14	
11.	10 + 5 = □		26.	15 = 9 + □	
12.	9 + 6 = □		27.	16 = 7 + □	
13.	9 + 1 = □		28.	10 + 4 = 9 + □	
14.	10 + 4 = □		29.	9 + 6 = 10 + □	
15.	9 + 5 = □		30.	□ + 6 = 7 + 9	

A

Number Correct:

Name ________________ Date ________________

*Write the missing number.

1.	9 + 2 = □		16.	4 + 8 = □	
2.	9 + 3 = □		17.	8 + 4 = □	
3.	9 + 5 = □		18.	7 + 4 = □	
4.	9 + 4 = □		19.	7 + 5 = □	
5.	8 + 2 = □		20.	7 + 6 = □	
6.	8 + 3 = □		21.	6 + 7 = □	
7.	8 + 5 = □		22.	9 + 9 = □	
8.	8 + 4 = □		23.	9 + □ = 18	
9.	9 + 4 = □		24.	□ + 4 = 13	
10.	8 + 5 = □		25.	□ + 4 = 12	
11.	9 + 5 = □		26.	12 = 3 + □	
12.	8 + 6 = □		27.	16 = 8 + □	
13.	9 + 6 = □		28.	9 + 4 = 8 + □	
14.	6 + 9 = □		29.	9 + 3 = 5 + □	
15.	9 + 6 = □		30.	□ + 7 = 8 + 6	

B

Number Correct:

Name ______________________ Date ____________

*Write the missing number.

1.	9 + 1 = □		16.	3 + 8 = □	
2.	9 + 2 = □		17.	8 + 3 = □	
3.	9 + 4 = □		18.	7 + 3 = □	
4.	9 + 3 = □		19.	7 + 4 = □	
5.	8 + 2 = □		20.	7 + 5 = □	
6.	8 + 3 = □		21.	5 + 7 = □	
7.	8 + 5 = □		22.	8 + 8 = □	
8.	8 + 4 = □		23.	8 + □ = 16	
9.	9 + 4 = □		24.	□ + 3 = 12	
10.	8 + 5 = □		25.	□ + 4 = 12	
11.	9 + 5 = □		26.	12 = 3 + □	
12.	8 + 7 = □		27.	14 = 7 + □	
13.	9 + 7 = □		28.	9 + 3 = 8 + □	
14.	7 + 9 = □		29.	9 + 3 = 5 + □	
15.	9 + 7 = □		30.	□ + 7 = 8 + 5	

ooooo ooooo

5-group row insert

A

Number Correct:

Name ______________________ Date ____________

*Write the missing number.

1.	10 - 9 = □		16.	10 - □ = 5	
2.	10 - 8 = □		17.	9 - □ = 5	
3.	10 - 6 = □		18.	8 - □ = 5	
4.	10 - 7 = □		19.	10 - □ = 3	
5.	10 - 6 = □		20.	9 - □ = 3	
6.	10 - 5 = □		21.	8 - □ = 3	
7.	10 - 6 = □		22.	□ - 6 = 4	
8.	10 - 4 = □		23.	□ - 6 = 3	
9.	10 - 3 = □		24.	□ - 6 = 2	
10.	10 - 7 = □		25.	10 - 4 = 9 - □	
11.	10 - 8 = □		26.	8 - 2 = 10 - □	
12.	10 - 2 = □		27.	8 - □ = 10 - 3	
13.	10 - 1 = □		28.	9 - □ = 10 - 3	
14.	10 - 9 = □		29.	10 - 4 = 9 - □	
15.	10 - 10 = □		30.	□ - 2 = 10 - 4	

B

Number Correct:

Name ______________________ Date ____________

*Write the missing number.

1.	10 - 8 = □		16.	10 - □ = 0	
2.	10 - 9 = □		17.	9 - □ = 0	
3.	10 - 8 = □		18.	8 - □ = 0	
4.	10 - 9 = □		19.	10 - □ = 1	
5.	10 - 7 = □		20.	9 - □ = 1	
6.	10 - 9 = □		21.	8 - □ = 1	
7.	10 - 8 = □		22.	□ - 5 = 5	
8.	10 - 7 = □		23.	□ - 5 = 4	
9.	10 - 3 = □		24.	□ - 5 = 3	
10.	10 - 7 = □		25.	10 - 8 = 9 - □	
11.	10 - 6 = □		26.	8 - 6 = 10 - □	
12.	10 - 4 = □		27.	8 - □ = 10 - 2	
13.	10 - 3 = □		28.	9 - □ = 10 - 2	
14.	10 - 7 = □		29.	10 - 3 = 9 - □	
15.	10 - 5 = □		30.	□ - 1 = 10 - 3	

A

Number Correct:

Name ______________________ Date ____________

*Write the missing number. Pay attention to the addition or subtraction sign.

1.	10 - 9 = □		16.	10 - 9 = □	
2.	1 + 2 = □		17.	11 - 9 = □	
3.	10 - 9 = □		18.	12 - 9 = □	
4.	1 + 3 = □		19.	15 - 9 = □	
5.	10 - 9 = □		20.	14 - 9 = □	
6.	1 + 1 = □		21.	13 - 9 = □	
7.	10 - 9 = □		22.	17 - 9 = □	
8.	1 + 2 = □		23.	18 - 9 = □	
9.	12 - 9 = □		24.	9 + □= 13	
10.	10 - 9 = □		25.	9 + □= 14	
11.	1 + 3 = □		26.	9 + □= 16	
12.	13 - 9 = □		27.	9 + □= 15	
13.	10 - 9 = □		28.	9 + □= 17	
14.	1 + 5 = □		29.	9 + □= 18	
15.	15 - 9 = □		30.	9 + □= 19	

B

Number Correct:

Name ______________________ Date ____________

*Write the missing number. Pay attention to the addition or subtraction sign.

1.	10 - 9 = □		16.	10 - 9 = □	
2.	1 + 1 = □		17.	11 - 9 = □	
3.	10 - 9 = □		18.	13 - 9 = □	
4.	1 + 2 = □		19.	14 - 9 = □	
5.	10 - 9 = □		20.	13 - 9 = □	
6.	1 + 3 = □		21.	12 - 9 = □	
7.	10 - 9 = □		22.	15 - 9 = □	
8.	1 + 4 = □		23.	16 - 9 = □	
9.	14 - 9 = □		24.	9 + □= 12	
10.	10 - 9 = □		25.	9 + □= 13	
11.	1 + 3 = □		26.	9 + □= 15	
12.	13 - 9 = □		27.	9 + □= 14	
13.	10 - 9 = □		28.	9 + □= 15	
14.	1 + 2 = □		29.	9 + □= 17	
15.	12 - 9 = □		30.	9 + □= 16	

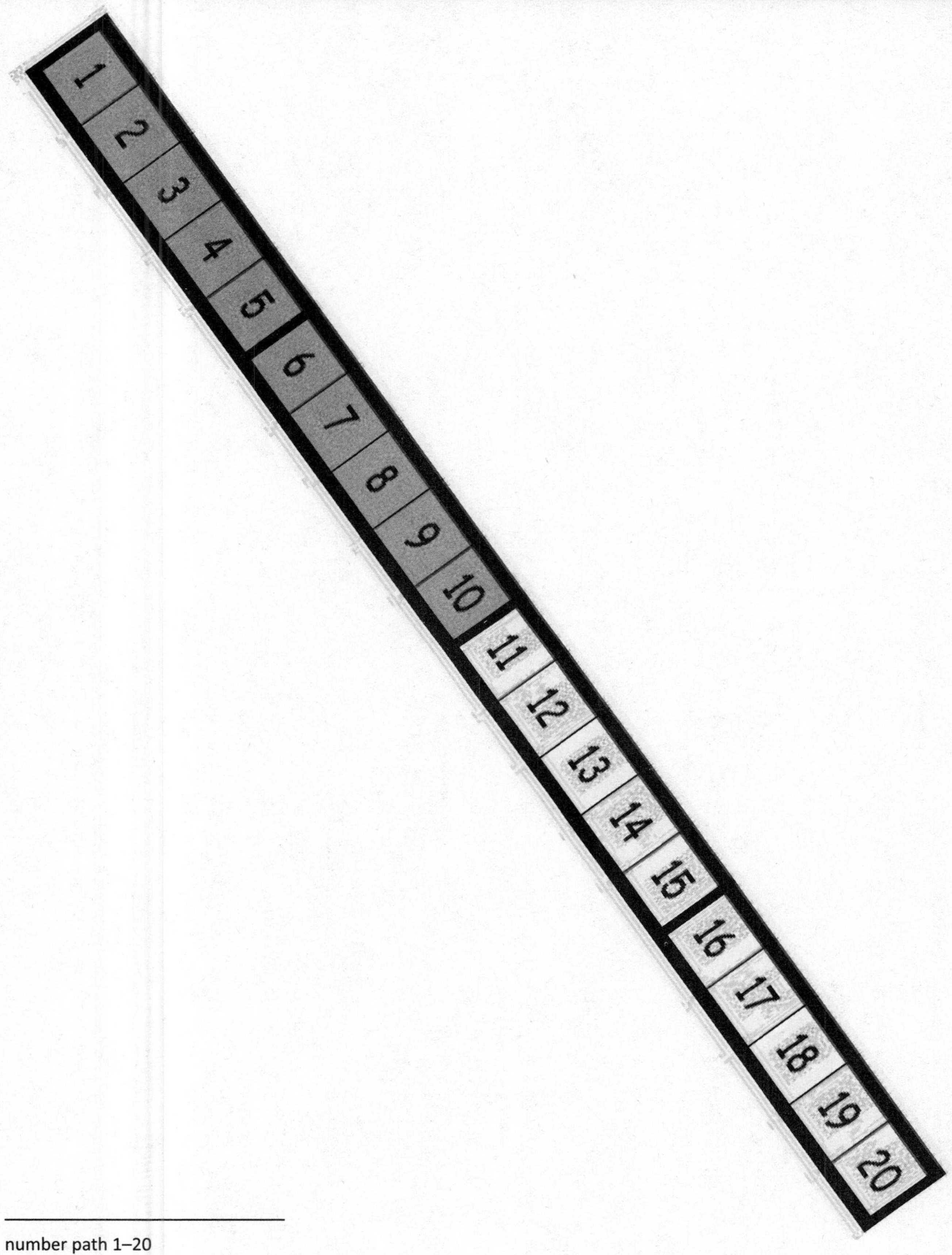

number path 1–20

A

Number Correct:

Name ______________________ Date ____________

*Write the missing number. Pay attention to the addition or subtraction sign.

1.	10 - 8 = □		16.	10 - 8 = □	
2.	2 + 2 = □		17.	11 - 8 = □	
3.	10 - 8 = □		18.	12 - 8 = □	
4.	2 + 3 = □		19.	15 - 8 = □	
5.	10 - 8 = □		20.	14 - 8 = □	
6.	2 + 4 = □		21.	13 - 8 = □	
7.	10 - 8 = □		22.	17 - 8 = □	
8.	2 + 1 = □		23.	18 - 8 = □	
9.	11 - 8 = □		24.	8 + □ = 11	
10.	10 - 8 = □		25.	8 + □ = 12	
11.	2 + 2 = □		26.	8 + □ = 15	
12.	12 - 8 = □		27.	8 + □ = 14	
13.	10 - 8 = □		28.	8 + □ = 16	
14.	2 + 5 = □		29.	8 + □ = 17	
15.	15 - 8 = □		30.	8 + □ = 18	

B

Number Correct:

Name ______________________ Date ____________

*Write the missing number. Pay attention to the addition or subtraction sign.

1.	10 - 8 = □		16.	10 - 8 = □	
2.	2 + 1 = □		17.	11 - 8 = □	
3.	10 - 8 = □		18.	13 - 8 = □	
4.	2 + 2 = □		19.	14 - 8 = □	
5.	10 - 8 = □		20.	13 - 8 = □	
6.	2 + 3 = □		21.	12 - 8 = □	
7.	10 - 8 = □		22.	15 - 8 = □	
8.	2 + 2 = □		23.	16 - 8 = □	
9.	12 - 8 = □		24.	8 + □ = 10	
10.	10 - 8 = □		25.	8 + □ = 11	
11.	2 + 3 = □		26.	8 + □ = 13	
12.	13 - 8 = □		27.	8 + □ = 12	
13.	10 - 8 = □		28.	8 + □ = 13	
14.	2 + 2 = □		29.	8 + □ = 15	
15.	12 - 8 = □		30.	8 + □ = 16	

A

Number Correct:

Name ______________________ Date ____________

*Write the missing number.

1.	10 - 9 = □		16.	12 - 7 = □	
2.	11 - 9 = □		17.	13 - 7 = □	
3.	13 - 9 = □		18.	14 - 7 = □	
4.	10 - 8 = □		19.	15 - 9 = □	
5.	11 - 8 = □		20.	15 - 8 = □	
6.	13 - 8 = □		21.	15 - 7 = □	
7.	10 - 7 = □		22.	17 - 7 = □	
8.	11 - 7 = □		23.	16 - 7 = □	
9.	13 - 7 = □		24.	17 - 7 = □	
10.	12 - 9 = □		25.	16 - □ = 9	
11.	13 - 9 = □		26.	16 - □ = 8	
12.	14 - 9 = □		27.	17 - □ = 8	
13.	12 - 8 = □		28.	17 - □ = 9	
14.	13 - 8 = □		29.	17 - □ = 16 - 8	
15.	14 - 8 = □		30.	□ - 7 = 17 - 8	

B

Number Correct:

Name ______________________ Date ____________

*Write the missing number.

1.	10 - 9 = □		16.	11 - 7 = □	
2.	11 - 9 = □		17.	12 - 7 = □	
3.	12 - 9 = □		18.	15 - 7 = □	
4.	10 - 8 = □		19.	15 - 9 = □	
5.	11 - 8 = □		20.	15 - 8 = □	
6.	12 - 8 = □		21.	15 - 7 = □	
7.	10 - 7 = □		22.	15 - 8 = □	
8.	11 - 7 = □		23.	16 - 8 = □	
9.	12 - 7 = □		24.	16 - 7 = □	
10.	11 - 9 = □		25.	16 - □ = 9	
11.	12 - 9 = □		26.	16 - □ = 8	
12.	15 - 9 = □		27.	16 - □ = 7	
13.	11 - 8 = □		28.	16 - □ = 9	
14.	12 - 8 = □		29.	16 - □ = 15 - 8	
15.	15 - 8 = □		30.	□ - 8 = 15 - 7	

A

Number Correct:

Name ______________________ Date ______________

*Write the missing number.

1.	2 + □ = 3		16.	2 + □ = 8	
2.	1 + □ = 3		17.	4 + □ = 8	
3.	□ + 1 = 3		18.	8 = □ + 6	
4.	□ + 2 = 4		19.	8 = 3 + □	
5.	3 + □ = 4		20.	□ + 3 = 9	
6.	1 + □ = 4		21.	2 + □ = 9	
7.	1 + □ = 5		22.	9 = □ + 1	
8.	4 + □ = 5		23.	9 = 4 + □	
9.	3 + □ = 5		24.	2 + 2 + □ = 9	
10.	3 + □ = 6		25.	2 + 2 + □ = 8	
11.	□ + 2 = 6		26.	3 + □ + 3 = 9	
12.	0 + □ = 6		27.	3 + □ + 2 = 9	
13.	1 + □ = 7		28.	5 + 3 = □ + 4	
14.	□ + 5 = 7		29.	□ + 4 = 1 + 5	
15.	□ + 4 = 7		30.	3 + □ = 2 + 6	

B

Number Correct:

Name ______________________ Date __________

*Write the missing number.

1.	1 + □ = 3		16.	3 + □ = 8	
2.	0 + □ = 3		17.	2 + □ = 8	
3.	□ + 3 = 3		18.	8 = □ + 1	
4.	□ + 2 = 4		19.	8 = 4 + □	
5.	3 + □ = 4		20.	□ + 2 = 9	
6.	4 + □ = 4		21.	4 + □ = 9	
7.	4 + □ = 5		22.	9 = □ + 5	
8.	1 + □ = 5		23.	9 = 6 + □	
9.	2 + □ = 5		24.	1 + 5 + □ = 9	
10.	4 + □ = 6		25.	3 + 2 + □ = 8	
11.	□ + 2 = 6		26.	2 + □ + 6 = 9	
12.	3 + □ = 6		27.	3 + □ + 4 = 9	
13.	3 + □ = 7		28.	5 + 4 = □ + 6	
14.	□ + 4 = 7		29.	□ + 3 = 6 + 2	
15.	□ + 5 = 7		30.	4 + □ = 2 + 7	

A

Number Correct:

Name ____________________ Date ____________

*Write the missing number.

1.	2 + □ = 3		16.	2 + □ = 8	
2.	1 + □ = 3		17.	4 + □ = 8	
3.	□ + 1 = 3		18.	8 = □ + 6	
4.	□ + 2 = 4		19.	8 = 3 + □	
5.	3 + □ = 4		20.	□ + 3 = 9	
6.	1 + □ = 4		21.	2 + □ = 9	
7.	1 + □ = 5		22.	9 = □ + 1	
8.	4 + □ = 5		23.	9 = 4 + □	
9.	3 + □ = 5		24.	2 + 2 + □ = 9	
10.	3 + □ = 6		25.	2 + 2 + □ = 8	
11.	□ + 2 = 6		26.	3 + □ + 3 = 9	
12.	0 + □ = 6		27.	3 + □ + 2 = 9	
13.	1 + □ = 7		28.	5 + 3 = □ + 4	
14.	□ + 5 = 7		29.	□ + 4 = 1 + 5	
15.	□ + 4 = 7		30.	3 + □ = 2 + 6	

B

Number Correct:

Name ______________________ Date ____________

*Write the missing number.

1.	$1 + \square = 3$		16.	$3 + \square = 8$	
2.	$0 + \square = 3$		17.	$2 + \square = 8$	
3.	$\square + 3 = 3$		18.	$8 = \square + 1$	
4.	$\square + 2 = 4$		19.	$8 = 4 + \square$	
5.	$3 + \square = 4$		20.	$\square + 2 = 9$	
6.	$4 + \square = 4$		21.	$4 + \square = 9$	
7.	$4 + \square = 5$		22.	$9 = \square + 5$	
8.	$1 + \square = 5$		23.	$9 = 6 + \square$	
9.	$2 + \square = 5$		24.	$1 + 5 + \square = 9$	
10.	$4 + \square = 6$		25.	$3 + 2 + \square = 8$	
11.	$\square + 2 = 6$		26.	$2 + \square + 6 = 9$	
12.	$3 + \square = 6$		27.	$3 + \square + 4 = 9$	
13.	$3 + \square = 7$		28.	$5 + 4 = \square + 6$	
14.	$\square + 4 = 7$		29.	$\square + 3 = 6 + 2$	
15.	$\square + 5 = 7$		30.	$4 + \square = 2 + 7$	

A

Number Correct:

Name ____________________ Date ____________

*Write the missing number.

1.	2 - □ = 1		16.	6 - □ = 2	
2.	2 - □ = 2		17.	6 - □ = 3	
3.	2 - □ = 0		18.	6 - □ = 4	
4.	3 - □ = 2		19.	7 - □ = 3	
5.	3 - □ = 1		20.	7 - □ = 2	
6.	3 - □ = 0		21.	7 - □ = 1	
7.	3 - □ = 3		22.	8 - □ = 2	
8.	4 - □ = 4		23.	8 - □ = 3	
9.	4 - □ = 3		24.	4 = 8 - □	
10.	4 - □ = 2		25.	2 = 9 - □	
11.	4 - □ = 1		26.	3 = 9 - □	
12.	5 - □ = 0		27.	4 = 9 - □	
13.	5 - □ = 1		28.	10 - 3 = 9 - □	
14.	5 - □ = 2		29.	9 - □ = 10 - 5	
15.	5 - □ = 3		30.	9 - □ = 10 - 6	

B

Number Correct:

Name ______________________ Date ____________

*Write the missing number.

1.	2 - □ = 2		16.	6 - □ = 3	
2.	2 - □ = 1		17.	6 - □ = 4	
3.	2 - □ = 0		18.	6 - □ = 5	
4.	3 - □ = 3		19.	7 - □ = 4	
5.	3 - □ = 2		20.	7 - □ = 3	
6.	3 - □ = 1		21.	7 - □ = 2	
7.	3 - □ = 0		22.	8 - □ = 3	
8.	4 - □ = 4		23.	8 - □ = 4	
9.	4 - □ = 3		24.	5 = 8 - □	
10.	4 - □ = 2		25.	3 = 9 - □	
11.	4 - □ = 1		26.	4 = 9 - □	
12.	5 - □ = 5		27.	5 = 9 - □	
13.	5 - □ = 4		28.	10 - 4 = 9 - □	
14.	5 - □ = 3		29.	9 - □ = 10 - 6	
15.	5 - □ = 2		30.	9 - □ = 10 - 5	

A

Number Correct:

Name ________________________ Date ____________

*Write the missing number.

1.	□ = 4 + 1		16.	7 + 3 = 4 + □	
2.	□ = 4 + 2		17.	6 + 4 = 5 + □	
3.	□ = 4 + 3		18.	5 + 5 = 6 + □	
4.	□ = 5 + 1		19.	5 + 3 = □ + 1	
5.	□ = 5 + 2		20.	5 + 4 = □ + 5	
6.	□ = 5 + 3		21.	4 + 5 = □ + 5	
7.	□ = 6 + 1		22.	2 + □ = 6 + 2	
8.	8 = 7 + □		23.	4 + □ = 5 + 3	
9.	9 = 8 + □		24.	□ + 4 = 5 + 2	
10.	9 = □ + 1		25.	□ + 6 = 4 + 3	
11.	9 = □ + 9		26.	4 + 2 = 1 + □	
12.	8 = □ + 1		27.	3 + 4 = □ + 2	
13.	□ = 7 + 1		28.	4 + 4 = 2 + □	
14.	10 = 8 + □		29.	3 + □ = 2 + 7	
15.	10 = □ + 8		30.	□ + 2 = 2 + 6	

B

Number Correct:

Name ______________________ Date ____________

*Write the missing number.

1.	$\square = 3 + 1$		16.	$5 + 5 = 4 + \square$	
2.	$\square = 3 + 2$		17.	$6 + 4 = 7 + \square$	
3.	$\square = 3 + 3$		18.	$3 + 7 = 8 + \square$	
4.	$\square = 4 + 1$		19.	$5 + 2 = \square + 1$	
5.	$\square = 4 + 2$		20.	$5 + 3 = \square + 5$	
6.	$\square = 4 + 3$		21.	$4 + 4 = \square + 4$	
7.	$\square = 5 + 1$		22.	$3 + \square = 6 + 3$	
8.	$8 = 1 + \square$		23.	$4 + \square = 5 + 4$	
9.	$9 = 1 + \square$		24.	$\square + 4 = 2 + 5$	
10.	$8 = \square + 7$		25.	$\square + 6 = 3 + 4$	
11.	$8 = \square + 8$		26.	$4 + 3 = 1 + \square$	
12.	$7 = \square + 1$		27.	$4 + 4 = \square + 2$	
13.	$\square = 6 + 1$		28.	$4 + 5 = 2 + \square$	
14.	$10 = 9 + \square$		29.	$3 + \square = 2 + 6$	
15.	$10 = \square + 9$		30.	$\square + 2 = 2 + 7$	

A

Number Correct:

Name ______________________ Date ____________

*Write the missing number.

1.	10 + 3 = □		16.	10 + □ = 11	
2.	10 + 2 = □		17.	10 + □ = 12	
3.	10 + 1 = □		18.	5 + □ = 15	
4.	1 + 10 = □		19.	4 + □ = 14	
5.	4 + 10 = □		20.	□ + 10 = 17	
6.	6 + 10 = □		21.	17 - □ = 7	
7.	10 + 7 = □		22.	16 - □ = 6	
8.	8 + 10 = □		23.	18 - □ = 8	
9.	12 - 10 = □		24.	□ - 10 = 8	
10.	11 - 10 = □		25.	□ - 10 = 9	
11.	10 - 10 = □		26.	1 + 1 + 10 = □	
12.	13 - 10 = □		27.	2 + 2 + 10 = □	
13.	14 - 10 = □		28.	2 + 3 + 10 = □	
14.	15 - 10 = □		29.	4 + □ + 3 = 17	
15.	18 - 10 = □		30.	□ + 5 + 10 = 18	

B

Number Correct:

Name ____________________ Date ____________

*Write the missing number.

1.	10 + 1 = □		16.	10 + □ = 10	
2.	10 + 2 = □		17.	10 + □ = 11	
3.	10 + 3 = □		18.	2 + □ = 12	
4.	4 + 10 = □		19.	3 + □ = 13	
5.	5 + 10 = □		20.	□ + 10 = 13	
6.	6 + 10 = □		21.	13 - □ = 3	
7.	10 + 8 = □		22.	14 - □ = 4	
8.	8 + 10 = □		23.	16 - □ = 6	
9.	10 - 10 = □		24.	□ - 10 = 6	
10.	11 - 10 = □		25.	□ - 10 = 8	
11.	12 - 10 = □		26.	2 + 1 + 10 = □	
12.	13 - 10 = □		27.	3 + 2 + 10 = □	
13.	15 - 10 = □		28.	2 + 3 + 10 = □	
14.	17 - 10 = □		29.	4 + □ + 4 = 18	
15.	19 - 10 = □		30.	□ + 6 + 10 = 19	

A

Number Correct:

Name ______________________ Date ____________

*Write the missing number.

1.	10 + 2 = □		16.	12 + 3 = □	
2.	2 + 1 = □		17.	13 + 3 = □	
3.	10 + 3 = □		18.	14 + 3 = □	
4.	4 + 10 = □		19.	13 + 5 = □	
5.	4 + 2 = □		20.	14 + 5 = □	
6.	6 + 10 = □		21.	15 + 5 = □	
7.	10 + 3 = □		22.	4 + 14 = □	
8.	3 + 3 = □		23.	4 + 15 = □	
9.	10 + 6 = □		24.	12 + □ = 14	
10.	2 + 1 = □		25.	12 + □ = 15	
11.	12 + 1 = □		26.	12 + □ = 16	
12.	2 + 2 = □		27.	□ + 4 = 16	
13.	12 + 2 = □		28.	5 + □ = 16	
14.	3 + 3 = □		29.	5 + □ = 26	
15.	13 + 3 = □		30.	4 + □ = 36	

B

Number Correct:

Name ______________________ Date ____________

*Write the missing number.

1.	10 + 1 = □		16.	12 + 2 = □	
2.	1 + 1 = □		17.	13 + 2 = □	
3.	10 + 2 = □		18.	14 + 2 = □	
4.	3 + 10 = □		19.	13 + 4 = □	
5.	3 + 2 = □		20.	14 + 4 = □	
6.	5 + 10 = □		21.	15 + 4 = □	
7.	10 + 2 = □		22.	5 + 14 = □	
8.	2 + 2 = □		23.	5 + 15 = □	
9.	10 + 4 = □		24.	11 + □ = 12	
10.	2 + 1 = □		25.	11 + □ = 13	
11.	12 + 1 = □		26.	11 + □ = 14	
12.	1 + 1 = □		27.	□ + 3 = 14	
13.	11 + 1 = □		28.	6 + □ = 19	
14.	3 + 2 = □		29.	6 + □ = 29	
15.	13 + 2 = □		30.	5 + □ = 39	

Grade 1
Module 3

A

Number Correct:

Name ______________________ Date ____________

*Write the missing number.

1.	3 - 3 = □		16.	13 - 1 = □	
2.	13 - 3 = □		17.	13 - 2 = □	
3.	3 - 2 = □		18.	14 - 3 = □	
4.	13 - 2 = □		19.	14 - 4 = □	
5.	4 - 2 = □		20.	14 - 10 = □	
6.	14 - 2 = □		21.	17 - 5 = □	
7.	4 - 3 = □		22.	17 - 6 = □	
8.	14 - 3 = □		23.	17 - 10 = □	
9.	14 - 10 = □		24.	8 - □ = 5	
10.	7 - 6 = □		25.	18 - □ = 15	
11.	17 - 6 = □		26.	18 - □ = 13	
12.	17 - 10 = □		27.	19 - □ = 12	
13.	6 - 3 = □		28.	□ - 2 = 17	
14.	16 - 3 = □		29.	17 - 3 = 16 - □	
15.	16 - 10 = □		30.	19 - 6 = □ - 5	

B

Number Correct:

Name ______________________ Date ____________

*Write the missing number.

1.	2 - 2 = □		16.	14 - 1 = □	
2.	12 - 2 = □		17.	14 - 2 = □	
3.	2 - 1 = □		18.	15 - 3 = □	
4.	12 - 1 = □		19.	15 - 4 = □	
5.	3 - 3 = □		20.	15 - 10 = □	
6.	13 - 3 = □		21.	18 - 5 = □	
7.	3 - 2 = □		22.	18 - 6 = □	
8.	13 - 2 = □		23.	18 - 10 = □	
9.	13 - 10 = □		24.	7 - □ = 5	
10.	6 - 5 = □		25.	17 - □ = 15	
11.	16 - 5 = □		26.	17 - □ = 13	
12.	16 - 10 = □		27.	19 - □ = 13	
13.	4 - 2 = □		28.	□ - 3 = 16	
14.	14 - 2 = □		29.	17 - 4 = 16 - □	
15.	14 - 10 = □		30.	19 - 7 = □ - 6	

A

Number Correct:

Name ______________________ Date ______________

*Write the missing number. Pay attention to the + and - signs.

1.	$5 + 2 = \square$		16.	$13 + 6 = \square$	
2.	$15 + 2 = \square$		17.	$3 + 16 = \square$	
3.	$2 + 5 = \square$		18.	$19 - 2 = \square$	
4.	$12 + 5 = \square$		19.	$19 - 7 = \square$	
5.	$7 - 2 = \square$		20.	$4 + 15 = \square$	
6.	$17 - 2 = \square$		21.	$14 + 5 = \square$	
7.	$7 - 5 = \square$		22.	$18 - 6 = \square$	
8.	$17 - 5 = \square$		23.	$18 - 2 = \square$	
9.	$4 + 3 = \square$		24.	$13 + \square = 19$	
10.	$14 + 3 = \square$		25.	$\square - 6 = 13$	
11.	$3 + 4 = \square$		26.	$14 + \square = 19$	
12.	$13 + 4 = \square$		27.	$\square - 4 = 15$	
13.	$7 - 4 = \square$		28.	$\square - 5 = 14$	
14.	$17 - 4 = \square$		29.	$13 + 4 = 19 - \square$	
15.	$17 - 3 = \square$		30.	$18 - 6 = \square + 3$	

B

Number Correct:

Name ______________________ Date ______________

*Write the missing number. Pay attention to the + and – signs.

1.	5 + 1 = □		16.	12 + 7 = □	
2.	15 + 1 = □		17.	2 + 17 = □	
3.	1 + 5 = □		18.	18 - 2 = □	
4.	11 + 5 = □		19.	18 - 6 = □	
5.	6 - 1 = □		20.	3 + 16 = □	
6.	16 - 1 = □		21.	13 + 6 = □	
7.	6 - 5 = □		22.	17 - 4 = □	
8.	16 - 5 = □		23.	17 - 3 = □	
9.	4 + 5 = □		24.	12 + □ = 18	
10.	14 + 5 = □		25.	□ - 6 = 12	
11.	5 + 4 = □		26.	13 + □ = 19	
12.	15 + 4 = □		27.	□ - 3 = 16	
13.	9 - 4 = □		28.	□ - 3 = 17	
14.	19 - 4 = □		29.	11 + 6 = 19 - □	
15.	19 - 5 = □		30.	19 - 5 = □ + 3	

A

Number Correct:

Name ______________________ Date ______________

*Write the missing number.

1.	17 - 1 = □		16.	19 - 9 = □	
2.	15 - 1 = □		17.	18 - 9 = □	
3.	19 - 1 = □		18.	11 - 9 = □	
4.	15 - 2 = □		19.	16 - 5 = □	
5.	17 - 2 = □		20.	15 - 5 = □	
6.	18 - 2 = □		21.	14 - 5 = □	
7.	18 - 3 = □		22.	12 - 5 = □	
8.	18 - 5 = □		23.	12 - 6 = □	
9.	17 - 5 = □		24.	14 - □ = 11	
10.	19 - 5 = □		25.	14 - □ = 10	
11.	17 - 7 = □		26.	14 - □ = 9	
12.	18 - 7 = □		27.	15 - □ = 9	
13.	19 - 7 = □		28.	□ - 7 = 9	
14.	19 - 2 = □		29.	19 - 5 = 16 - □	
15.	19 - 7 = □		30.	15 - 8 = □ - 9	

B

Number Correct:

Name ____________________ Date __________

*Write the missing number.

1.	16 - 1 = □		16.	19 - 9 = □	
2.	14 - 1 = □		17.	18 - 9 = □	
3.	18 - 1 = □		18.	12 - 9 = □	
4.	19 - 2 = □		19.	19 - 8 = □	
5.	17 - 2 = □		20.	18 - 8 = □	
6.	15 - 2 = □		21.	17 - 8 = □	
7.	15 - 3 = □		22.	14 - 5 = □	
8.	17 - 5 = □		23.	13 - 5 = □	
9.	19 - 5 = □		24.	12 - □ = 7	
10.	16 - 5 = □		25.	16 - □ = 10	
11.	16 - 6 = □		26.	16 - □ = 9	
12.	19 - 6 = □		27.	17 - □ = 9	
13.	17 - 6 = □		28.	□ - 7 = 9	
14.	17 - 1 = □		29.	19 - 4 = 17 - □	
15.	17 - 6 = □		30.	16 - 8 = □ - 9	

A

Number Correct:

Name ____________________ Date ____________

*Write the missing number.

1.	17 + 1 = □		16.	11 + 9 = □	
2.	15 + 1 = □		17.	10 + 9 = □	
3.	18 + 1 = □		18.	9 + 9 = □	
4.	15 + 2 = □		19.	7 + 9 = □	
5.	17 + 2 = □		20.	8 + 8 = □	
6.	18 + 2 = □		21.	7 + 8 = □	
7.	15 + 3 = □		22.	8 + 5 = □	
8.	5 + 13 = □		23.	11 + 8 = □	
9.	15 + 2 = □		24.	12 + □ = 17	
10.	5 + 12 = □		25.	14 + □ = 17	
11.	12 + 4 = □		26.	8 + □ = 17	
12.	13 + 4 = □		27.	□ + 7 = 16	
13.	3 + 14 = □		28.	□ + 7 = 15	
14.	17 + 2 = □		29.	9 + 5 = 10 + □	
15.	12 + 7 = □		30.	7 + 8 = □ + 9	

B

Number Correct:

Name ______________________ Date ______________

*Write the missing number.

1.	14 + 1 = □		16.	11 + 9 = □	
2.	16 + 1 = □		17.	10 + 9 = □	
3.	17 + 1 = □		18.	8 + 9 = □	
4.	11 + 2 = □		19.	9 + 9 = □	
5.	15 + 2 = □		20.	9 + 8 = □	
6.	17 + 2 = □		21.	8 + 8 = □	
7.	15 + 4 = □		22.	8 + 5 = □	
8.	4 + 15 = □		23.	11 + 7 = □	
9.	15 + 3 = □		24.	12 + □ = 18	
10.	5 + 13 = □		25.	14 + □ = 18	
11.	13 + 4 = □		26.	8 + □ = 18	
12.	14 + 4 = □		27.	□ + 5 = 14	
13.	4 + 14 = □		28.	□ + 6 = 15	
14.	16 + 3 = □		29.	9 + 6 = 10 + □	
15.	13 + 6 = □		30.	6 + 7 = □ + 9	

A

Number Correct:

Name ______________________________ Date ______________

*Write the missing number.

1.	17 + 1 = □		16.	11 + 9 = □	
2.	15 + 1 = □		17.	10 + 9 = □	
3.	18 + 1 = □		18.	9 + 9 = □	
4.	15 + 2 = □		19.	7 + 9 = □	
5.	17 + 2 = □		20.	8 + 8 = □	
6.	18 + 2 = □		21.	7 + 8 = □	
7.	15 + 3 = □		22.	8 + 5 = □	
8.	5 + 13 = □		23.	11 + 8 = □	
9.	15 + 2 = □		24.	12 + □ = 17	
10.	5 + 12 = □		25.	14 + □ = 17	
11.	12 + 4 = □		26.	8 + □ = 17	
12.	13 + 4 = □		27.	□ + 7 = 16	
13.	3 + 14 = □		28.	□ + 7 = 15	
14.	17 + 2 = □		29.	9 + 5 = 10 + □	
15.	12 + 7 = □		30.	7 + 8 = □ + 9	

B

Number Correct:

Name ____________________ Date ____________

*Write the missing number.

1.	14 + 1 = □		16.	11 + 9 = □	
2.	16 + 1 = □		17.	10 + 9 = □	
3.	17 + 1 = □		18.	8 + 9 = □	
4.	11 + 2 = □		19.	9 + 9 = □	
5.	15 + 2 = □		20.	9 + 8 = □	
6.	17 + 2 = □		21.	8 + 8 = □	
7.	15 + 4 = □		22.	8 + 5 = □	
8.	4 + 15 = □		23.	11 + 7 = □	
9.	15 + 3 = □		24.	12 + □ = 18	
10.	5 + 13 = □		25.	14 + □ = 18	
11.	13 + 4 = □		26.	8 + □ = 18	
12.	14 + 4 = □		27.	□ + 5 = 14	
13.	4 + 14 = □		28.	□ + 6 = 15	
14.	16 + 3 = □		29.	9 + 6 = 10 + □	
15.	13 + 6 = □		30.	6 + 7 = □ + 9	

A

Number Correct:

Name ____________________ Date ____________

*Write the missing number.

1.	17 - 1 = □		16.	19 - 9 = □	
2.	15 - 1 = □		17.	18 - 9 = □	
3.	19 - 1 = □		18.	11 - 9 = □	
4.	15 - 2 = □		19.	16 - 5 = □	
5.	17 - 2 = □		20.	15 - 5 = □	
6.	18 - 2 = □		21.	14 - 5 = □	
7.	18 - 3 = □		22.	12 - 5 = □	
8.	18 - 5 = □		23.	12 - 6 = □	
9.	17 - 5 = □		24.	14 - □ = 11	
10.	19 - 5 = □		25.	14 - □ = 10	
11.	17 - 7 = □		26.	14 - □ = 9	
12.	18 - 7 = □		27.	15 - □ = 9	
13.	19 - 7 = □		28.	□ - 7 = 9	
14.	19 - 2 = □		29.	19 - 5 = 16 - □	
15.	19 - 7 = □		30.	15 - 8 = □ - 9	

B

Number Correct:

Name ______________________ Date ____________

*Write the missing number.

1.	16 - 1 = □		16.	19 - 9 = □	
2.	14 - 1 = □		17.	18 - 9 = □	
3.	18 - 1 = □		18.	12 - 9 = □	
4.	19 - 2 = □		19.	19 - 8 = □	
5.	17 - 2 = □		20.	18 - 8 = □	
6.	15 - 2 = □		21.	17 - 8 = □	
7.	15 - 3 = □		22.	14 - 5 = □	
8.	17 - 5 = □		23.	13 - 5 = □	
9.	19 - 5 = □		24.	12 - □ = 7	
10.	16 - 5 = □		25.	16 - □ = 10	
11.	16 - 6 = □		26.	16 - □ = 9	
12.	19 - 6 = □		27.	17 - □ = 9	
13.	17 - 6 = □		28.	□ - 7 = 9	
14.	17 - 1 = □		29.	19 - 4 = 17 - □	
15.	17 - 6 = □		30.	16 - 8 = □ - 9	

A

Number Correct:

Name ____________________ Date ____________

*Write the missing number.

1.	9 + 1 + 3 = □		16.	6 + 3 + 8 = □	
2.	9 + 2 + 1 = □		17.	5 + 9 + 4 = □	
3.	5 + 5 + 3 = □		18.	3 + 12 + 4 = □	
4.	5 + 2 + 5 = □		19.	3 + 11 + 5 = □	
5.	4 + 5 + 5 = □		20.	5 + 6 + 7 = □	
6.	8 + 2 + 4 = □		21.	2 + 6 + 3 = □	
7.	8 + 3 + 2 = □		22.	3 + 2 + 13 = □	
8.	12 + 2 + 2 = □		23.	3 + 13 + 3 = □	
9.	3 + 3 + 12 = □		24.	9 + 1 + □ = 14	
10.	4 + 4 + 5 = □		25.	8 + 4 + □ = 16	
11.	2 + 15 + 2 = □		26.	□ + 8 + 6 = 19	
12.	7 + 3 + 3 = □		27.	2 + □ + 7 = 18	
13.	1 + 17 + 1 = □		28.	2 + 2 + □ = 18	
14.	14 + 2 + 2 = □		29.	19 = 6 + □ + 9	
15.	4 + 12 + 4 = □		30.	18 = 7 + □ + 6	

B

Number Correct:

Name ______________________ Date ____________

*Write the missing number.

1.	9 + 1 + 2 = □		16.	6 + 3 + 9 = □	
2.	9 + 4 + 1 = □		17.	4 + 9 + 2 = □	
3.	5 + 5 + 1 = □		18.	2 + 12 + 4 = □	
4.	5 + 3 + 5 = □		19.	2 + 11 + 5 = □	
5.	4 + 5 + 5 = □		20.	6 + 6 + 7 = □	
6.	8 + 2 + 2 = □		21.	2 + 6 + 5 = □	
7.	8 + 3 + 2 = □		22.	3 + 3 + 13 = □	
8.	11 + 1 + 1 = □		23.	3 + 14 + 3 = □	
9.	2 + 2 + 14 = □		24.	9 + 1 + □ = 13	
10.	4 + 4 + 4 = □		25.	8 + 4 + □ = 15	
11.	2 + 13 + 2 = □		26.	□ + 8 + 6 = 18	
12.	6 + 3 + 3 = □		27.	2 + □ + 6 = 18	
13.	1 + 15 + 1 = □		28.	2 + 5 + □ = 18	
14.	15 + 2 + 2 = □		29.	19 = 5 + □ + 9	
15.	3 + 14 + 3 = □		30.	19 = 7 + □ + 6	

Credits

Great Minds® has made every effort to obtain permission for the reprinting of all copyrighted material. If any owner of copyrighted material is not acknowledged herein, please contact Great Minds for proper acknowledgment in all future editions and reprints of this module.